Modern Regression Techniques Using R

Modern Regression Techniques Using R

A Practical Guide for Students and Researchers

Daniel B. Wright and Kamala London

Los Angeles | London | New Delhi
Singapore | Washington DC

First published 2009
Reprinted 2010, 2011, 2012

SAGE Publications Ltd
1 Oliver's Yard
55 City Road
London EC1Y 1SP

SAGE Publications Inc.
2455 Teller Road
Thousand Oaks, California 91320

SAGE Publications India Pvt Ltd
B 1/I 1 Mohan Cooperative Industrial Area
Mathura Road
New Delhi 110 044

SAGE Publications Asia-Pacific Pte Ltd
33 Pekin Street #02-01
Far East Square
Singapore 048763

Library of Congress Control Number: 2008926086

British Library Cataloguing in Publication data

A catalogue record for this book is available from
the British Library

ISBN 978-1-84787-902-8
ISBN 978-1-84787-903-5 (pbk)

Typeset by CEPHA Imaging Pvt. Ltd., Bangalore, India
Printed in Great Britain by the MPG Books Group
Printed on paper from sustainable resources

Contents

Preface

In this book we introduce several useful extensions to the basic regression model, without too much mathematics, but with several pictures and some of the basic references. Not all possible extensions are covered, but we chose a set that we think is particularly useful for psychology. We will use the freeware package R so a secondary purpose of this book is to introduce some of the facilities in R (R Development Core Team, 2008). It works like syntax in many of the other statistics packages like SPSS (which seems the most popular package in psychology, so we refer to it occasionally for comparison purposes), but it is more flexible and has more procedures. Once you get used to it, we hope that you will find it is easier than its competitors. It is free so we know you will like the price! While we provide a brief introduction to R, we also provide links to useful books and websites.

This book is divided into ten chapters. First, we explain the most basic basics of R, but point readers to where they can find more details. Next, we give an overview of what we call the basic regression and then briefly describe each of the extensions. Then, we go through the seven extensions, and finish with a conclusion. Each of these chapters includes a description and then goes through the analysis of some data.

This document grew out of a regression workshop to the Legal Psychology group at Florida International University in 2006, when Dan was on sabbatical there (and where he is now permanently), and was the basis for a talk and poster at the SARMAC 2007 conference at Bates College, Maine. It was also the basis of *Modern Statistical Methods*, a graduate course at University of Sussex. Many thanks to all those who provided comments!

All of the royalties from this book go to the American Partnership for Eosinophilic Disorders (www.apfed.com).

See the website for more information.

Happy regressing,

Dan Wright and Kami London

GETTING THE MOST OUT OF THIS BOOK

An important part of using this book is conducting analyses in R. R is freely available on the web (http://cran.r-project.org/). You should, at least to begin with, be on a computer with internet access. R is updated frequently and some minor aspects are changed each time. This book was prepared with R2.4 thru R2.7. R is part of the Free Software Foundation (http://www.fsf.org/), which promotes, well, the name makes it pretty obvious what they promote.

In this book R commands are written in **dark bold Courier** and R output is in gray Courier. There is a glossary at the back of the book which provides brief descriptions of all the commands/functions used in this book, so if something is unfamiliar look there first. If you want to know more about the function use the online help facility within R. To do this you should use the **help** function. For example, for the function **mean**, type either **help(mean)** or **?mean**.

We have adopted an example-based approach. Most of the data come from real research papers. The examples were chosen because we hope that they will be of interest to most working in the social and behavioral sciences, and also because we were able to access the data. By providing examples, we hope you can match your own research needs onto these examples. The data for all these examples and the corresponding code are on http://www.sagepub.co.uk/wrightandlondon.

There are many books that cover conducting statistics in R. A list of some can be found at:

http://www.r-project.org/doc/bib/R-books.html

One of our favorites is:

Crawley, M. J. (2005). *Statistics: An introduction using R*. Chichester, UK: Wiley.

This is not written specifically for social scientists, but his clarity is excellent. He also has written *The R book* (2007) which is excellent for a much more detailed treatment at 950 pages.

More detailed readings are given at the end of each chapter. We assume that everybody has studied some statistics, perhaps one semester of psychology-graduate-statistics, and so understands the basics of the standard linear regression (covered briefly in Chapter 2). There are several good background books for statistics, but one stands out above all others for having the most modest authors:

Wright, D. B. & London, K. (2009). *First (and second) steps in statistics (2nd)*. London: Sage Publications.

NOTE

Microsoft Word and many other 'high level' word processing packages change some characters (including " and ') to other characters (like " and '), which are not read by R. Therefore, if using one of these word processing packages we recommend turning off several of the facilities that automatically change characters from those you type. If copying code from websites, sometimes line breaks are lost, so you need to be careful with this. If you are copying and pasting commands, it may be easier to save them in Notepad or some other 'low level' word processing package. The text editor Tinn-R is designed for R and can be downloaded from http://www.sciviews.org/Tinn-R/ and http://sourceforge.net/projects/tinn-r.

1

Very brief introduction to R

Learning objectives

1. Learning some of the basic R concepts: functions, objects, assigning, packages, mirrors, CRAN, and how to read data and access packages.
2. Statistical concepts reinforced are looking at data, transforming data, and there is detailed discussion of *skewness*.
3. We introduce you to the *bootstrap*, which will be used for several examples in this book.

R was developed as a free alternative to the powerful statistics language/program S/S-Plus. R and S-Plus are similar, and many of the procedures written in one will run in the other, but R is free and S-Plus is a commercial product. R is rapidly increasing in popularity. When statisticians develop procedures they often write R functions so that others can use them. Several books for learning to use R are listed at the end of this chapter.

R AND THE INTERNET

When using R it is useful to have an internet connection. Figure 1.1 shows a schematic of how the R system can be considered. From your computer you download R from the internet onto your computer so that you can later use the software without being on the internet. The program is available both from the R home page and from one of the CRAN (Comprehensive R Archive Network) mirror sites. Mirror sites, or mirrors, are sites that are supposed to be exactly the same as the main CRAN site. This means that when people download files they do not all have to do it from the same server. This makes downloading faster.

To begin using R you have to download it from one of the many R mirror sites: http://cran.r-project.org/mirrors.html. If an entire class is downloading information you should use different mirrors. Press the Windows, the Mac, or the Linux button in the 'precompiled binary distribution' box. Press 'base' and then run the setup program. Follow the wizard's instructions. This gives you both a very powerful statistics language and statistics package. This will allow you to do most of the statistics that you

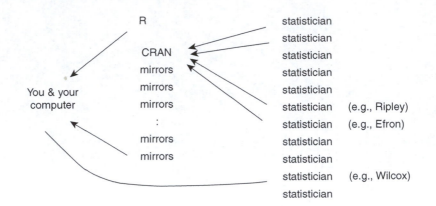

Figure 1.1   A schematic of the R-system

would want, but not all. Statisticians write their own packages for specialist purposes. Some submit these to CRAN so that others can use them. When a package is sent to CRAN it gets copied onto all the mirror sites. You can then download packages from there. For example, there is a package called **foreign** (R core members et al., 2008) that allows you to read data from other statistics programs directly into R. If you type:

```
install.packages("foreign")
```

a window like Figure 1.2 opens. We chose the server in Michigan (USA (MI)) since that is close to where we are preparing this chapter (in Ohio). This should install the package onto your computer. The package is now on your computer so you may access it in the future from this computer even if you are not connected to the internet, assuming it is not erased. However, because authors update their packages frequently, it is worth reinstalling packages relatively regularly.

If the mirror is not perfectly up-to-date or if you are not connected to the internet it may not install. You may also get some warning messages.

Although the package is now on your computer (in a folder in the R directory) it is not active. To make it active type:

```
library("foreign")
```

Now you have access to a large number of functions that are used to import and export data between R and other statistical packages.

Some packages will have been installed when you downloaded R and you will just need to load these. You will need to download others from CRAN. Some statisticians, like Rand Wilcox, keep their packages on their own web page. In the case of Wilcox, he has written a book that effectively acts as both a manual and teaching resource for his functions. We will use some of his code in one of the examples and his code can be accessed from the web using the **source** function. We have only written 'statistician' on the right of Figure 1.1, but there are people from other disciplines (like computer science and psychology) who write packages for R. We did not include them because it is primarily statisticians doing the writing.

CRAN mirror

Japan (Tokyo)
Japan (Tsukuba)
Korea
Netherlands (Amsterdam 1)
Netherlands (Amsterdam 2)
Norway
Poland (Lublin)
Poland (Wroclaw)
Portugal
Slovenia (Besnica)
Slovenia (Ljubljana)
South Africa (Cape Town)
South Africa (Grahamstown)
Spain (Alicante)
Spain (Madrid)
Sweden
Switzerland (Zuerich)
Switzerland (Bern 1)
Switzerland (Bern 2)
Turkey
Taiwan (Taichung)
Taiwan (Taipeh)
UK (Bristol)
UK (London)
USA (CA 1)
USA (CA 2)
USA (CA 3)
USA (IA)
USA (IL)
USA (MI)
USA (MO)
USA (NC)
USA (PA 1)
USA (PA 2)
USA (TX)
USA (WA)

OK   Cancel

Figure 1.2   The window to choose a mirror site

Most of the introductory R books (a long list of them is on http://www.r-project.org/doc/bib/R-books.html; our favorite introduction is Crawley, 2005) go through how to use R like a calculator. Type **6+4**, or something like that, and see what happens. These books also go through the different data formats. A good (short and also free) book is available on http://cran.r-project.org/doc/manuals/R-intro.pdf.

If you have not already opened R, it would be good to open up it now because we will be telling you to type things in throughout the rest of this chapter.

FUNCTIONS AND OBJECTS

R works by applying *functions* to *objects*. To illustrate functions and objects we will show how to calculate the mean of four numbers. We will use the function **mean** to do this, but first we have to create a variable. A variable is a set of several similar objects. We can create a variable by *assigning* a list of values to a variable name. To make assignments you use the **<-** or **->** characters. So the following (and type this into R yourself) creates a variable **scores** that has four numbers (5, 6, 7 and 8).

```
scores <- c(5,6,7,8)
```

Another function, **c**, which stands for concatenate, is frequently used in R. This function tells R that these four numbers are a set. If we type:

```
scores
```

we get:

```
[1] 5 6 7 8
```

scores is a numeric variable, which is a type of object. R is case sensitive so typing **Scores** would have produced an error. An alternative way to write a sequence of numbers is with a colon **:**, such that

```
5:8
```

also produces the sequence

```
[1] 5 6 7 8
```

There is also a sequence function, **seq**, which can be used:

```
seq(5,8)
```

```
[1] 5 6 7 8
```

This function is useful if you want more complex sequences. If you type:

```
seq(10,30,5)
```

you get the sequence from 10 to 30 in steps of 5 units:

```
[1] 10 15 20 25 30
```

In R everything is an object, including variables. We can apply the function **mean** to this type of object (i.e., to numeric variables) by typing:

```
mean(scores)
```

and we get:

```
[1]  6.5
```

The `[1]` in front of `6.5` is because some functions produce several pieces of information so their parts are labeled. You can use functions in creating new variables. For example, you may want to have a variable for how far away each value is from the mean of the variable. The following command does this:

`residscores <- scores - mean(scores)`

Typing

`residscores`

produces

```
[1]  -1.5 -0.5 0.5 1.5
```

Functions in R work with certain types of objects. While you can take a mean of four numbers, you cannot take a mean of four people's names. Names (or string values) need to be placed in quotation marks so that R does not think they are objects that it should be able to find. The following creates a variable of four people's names and shows that the function **mean** does not work in this instance.

```
Simpsons <- c("Homer", "Marge", "Lisa", "Bart")
mean(Simpsons)
```

```
[1] NA
Warning message:
argument is not numeric or logical: returning NA in:
mean.default(Simpsons)
```

The function **c** works also with strings. Thus, when Maggie arrived at the Simpson household, the glorious event of childbirth could be written in R as:

`Simpsons <- c(Simpsons,"Maggie")`

We can identify each member of the Simpsons by writing the variable name with an index. So, to identify the fourth member of the Simpsons, we write:

`Simpsons[4]`

```
[1] "Bart"
```

When a function is applied to an object it can create another object, which can then be used in other functions. The basic regression function is **lm**. It is applied to some data objects (a response variable and some predictor variables) and a **lm.object** is created. This object can then be used in other functions, like **plot**, **summary**, and **anova**.

These are illustrated in several examples throughout this book. Many R functions are intelligent. For example, the **plot** function works differently dependent upon what type of object is placed within it, and we will see this as we go through this book.

Example 1 – Chile heat and length

- Data: Downloaded from www.pepperjoe.com, May 2006
- Research question: Are the length and heat of a chile related?
- Purpose: Showing how to load data from a text file and from SPSS, how to access packages, and run some preliminary statistics and graphs. The statistical concepts are skewness, transformations, and simple linear regression.

The basic methods in R to read data assume that the data are in a text file in a table-like form. Figure 1.3 shows data in this format in a Notepad file. It is important to have data in a text (or ASCII) format, so if your data are in a word processing file you have to use the SAVE AS option. If the data are in text format and in a table like Figure 1.3, then the command:

```
newdata <- read.table("filename")
```

assigns the data in "filename" into the object **newdata**. When using **read.table** the **"filename"** may either be a file on your computer or one on the web. For the function used in this example, **read.spss**, the data must be saved on your computer. The characters **<-** assign things to each other, and can be used in either direction so **x <- 6+4** and **6+4 -> x** both assign the value 10 to **x**. Often you will have data in another package and need to import it into R. The package **foreign** allows data to be read from other packages, including: SPSS, Minitab, SYSTAT, SAS, and Stata. Since it seems SPSS is the most popular among academic psychologists we will assume that you want to import the data from SPSS.

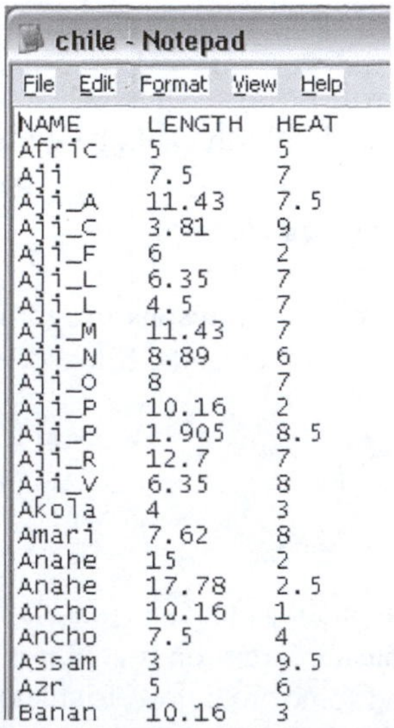

```
chile - Notepad

File  Edit  Format  View  Help
NAME      LENGTH    HEAT
Afric     5         5
Aji       7.5       7
Aji_A     11.43     7.5
Aji_C     3.81      9
Aji_F     6         2
Aji_L     6.35      7
Aji_L     4.5       7
Aji_M     11.43     7
Aji_N     8.89      6
Aji_O     8         7
Aji_P     10.16     2
Aji_P     1.905     8.5
Aji_R     12.7      7
Aji_V     6.35      8
Akola     4         3
Amari     7.62      8
Anahe     15        2
Anahe     17.78     2.5
Ancho     10.16     1
Ancho     7.5       4
Assam     3         9.5
Azr       5         6
Banan     10.16     3
```

Figure 1.3 A text file for the chile data set as shown within Notepad

PEPPER JOE'S
HOT PEPPER HEAT SCALE

10 NUCLEAR!	Caribbean - Golden Habanero - Org.Habanero
9.5	Brown Congo - Jamaican - Fatalli
9 TORRID	Hot Lemon - Firecracker
8.5	Thai Sun - Tabasco
8 Real Hot	Goat - Chicken Heart
7	Charleston - Peter - Fluorescent
6 FAIRLY HOT	Mushroom - Jalapeno
5	Bolivian Rainbow - Ancho
4 SORT OF HOT	Anaheim - Big Jim - Pepperoncini
3	Pasilla
2 FOR SISSYS	
1	

Figure 1.4   The heat scale. From www.pepperjoe.com/about/heatscale.html

The data in this example examine the relationship between the length of a chile and its heat. **LENGTH** is in cm and **HEAT** is in Pepper Joe's Hot Pepper Heat Scale shown in Figure 1.4 (in technical papers the Scoville scale is used, but they don't have a smiling chile with a shovel waving at you, see www.pepperjoe.com/about/heatscale.html).

The **library** function loads the package **foreign** so that it can be used. Note that \\ are used rather than \ in the **read.spss** function (and the **read.table** function used below). The **attach** command makes the data file the active data file, overwriting any other variables that may have the same names. The following commands read data from c:\temp\chile.sav. The data are available on the book's web page (on www.sagepub.co.uk/wrightandlondon) so this file will need to be copied into your c:\temp folder (if you do not have a c:\temp folder copy it elsewhere and change the code below accordingly).

```
library("foreign")
chile <- read.spss("c:\\temp\\chile.sav")
attach(chile)
```

You may get a warning, but this particular warning should not a problem; these data are read accurately. You do not need to have SPSS on your computer to read SPSS data files into R.

The data are also stored on this book's web page as a text file. They can be accessed by:

```
chile <- read.table
   ("http://www.sagepub.co.uk//wrightandlondon//chile.dat",
   header=T)
```

This command takes up multiple lines to print and requires that you very carefully write the web page each time. So that you do not have to write a command that is multiple lines, it is worth assigning this book's web page to the object **webreg** so that it need not be typed each time.

```
webreg <- "http://www.sagepub.co.uk//wrightandlondon//"
```

To write the full web address we have to **paste** together the web page and the file name.

```
paste(webreg,"chile.dat",sep="")
```

```
[1] "http://www.sagepub.co.uk//wrightandlondon//chile.sav"
```

The default for the **paste** function is to have one space between each of the objects that are pasted together. Having a space in a web address will cause the web address not to be recognized, so we have to tell R that there should be no separation between the objects. The option **sep=""** tells the computer this. If we wrote **sep=","** it would have placed a comma between each part.

Now we can write:

```
chile <- read.table(paste(webreg,"chile.dat",sep=""),
    header=T)
```

The **header=T** means that the first line in the data file has the variable names (see Figure 1.3). The **T** stands for **TRUE** and it can be written as **header=TRUE**. Alternative methods for reading data from the book's web page are listed in the box below. If you plan to use this book while disconnected from the internet, there are instructions on the web page for downloading all the data, functions, and syntax to your hard drive (or memory stick).

There are many ways to access data from the book's web page and three are presented here. The first is to write out the web page within the **read.table** command:

```
chile <- read.table
    ("http://www.sagepub.co.uk//wrightandlondon//chile.dat",
    header=T)
```

The problem is that this takes a few lines and requires careful typing.

The second is to first assign the web page to an object, like **webreg**, and then use the shorter command.

```
webreg <- "http://www.sagepub.co.uk//wrightandlondon//"
chile <- read.table(paste(webreg,"chile.dat",sep=""),
    header=T)
```

This still requires assigning the web address to **webreg** every time you turn on R and want to access the book's web page, but at least the first line can be copied and pasted from other exercises.

The final option, which will make life easier for people who are fairly comfortable with computers, is to add the line

```
webreg <- "http://www.sagepub.co.uk//wrightandlondon//"
```

into the file called Rprofile.site which was installed when you installed R. Use the search facility on your computer to find it. If you add this line then everytime you start R this assignment will be made. Then you would only need to type:

```
chile <- read.table(paste(webreg,"chile.dat",sep=""),
   header=T)
```

Because this final option requires more computer knowledge than we are assuming, the second option will be used throughout this book.

After you import the data set (and name it, say **chile**), then type:

```
attach(chile)
```

This attaches the data set to your working environment. 'Attaching to your working environment' is the technical jargon that means you can now access all these variables by just typing the variable names. To find out the variable names type:

```
names(chile)
```

```
[1] "NAME"  "LENGTH"  "HEAT"
```

These are the same as in Figure 1.3. While you can access a variable without the file being attached by typing:

```
chile$HEAT
```

this can get cumbersome and is not useful if you are working with only one data set at a time. It is easier not having to re-type the name of the data set each time. Therefore, in this book we will always **attach** the data set we are working with. If you are doing more advanced statistics where you are looking at several data sets, we recommend not attaching the data but accessing them within each function or using the **with** function (see Crawley, 2007). However, for most psychologists' needs this would add further complications.

When R reads SPSS files, all the variable names are UPPERCASE. This is how SPSS stores them internally because SPSS is not case sensitive. R is case sensitive so:

```
heat
```

produces the word NULL. This means there is no variable **heat**. The variable is **HEAT**.

If you want to find out how many cases are in the variable **LENGTH**, you need to ask for the **length** of **LENGTH**:

```
length(LENGTH)
```

length is an R function, **LENGTH** is a variable in the chile dataset. Be careful not to assign **length <- LENGTH**, because, while R will try to figure out what you want, this can lead to errors.

> We are now going to describe *skewness* in more depth than is typical in introductory statistics courses. We do this for two reasons. First, it emphasizes the importance of looking at your variables and describing their distributions. This is something that the American Psychological Association states should be done (Wilkinson et al., 1999). Second, it provides a good way to focus on how well different transformations work. Statisticians stress the importance of transformations (e.g., Mosteller & Tukey, 1977), but this stress is often absent from psychology texts.

We now look at the histogram of this variable. This can be done with **hist(LENGTH)** and the result is shown in the left panel of Figure 1.5. It looks fairly skewed. There are several different definitions of skewness (Groeneveld & Meeden, 1984) and none are in the base package R, so you have to load another package with this command, the two most commonly used being **e1071** (Dimitriadou et al., 2008) and **fBasics** (Harrell et al., 2007). This loads a function, **skewness**, which is the most common measure of skewness.[1] It is:

$$skewness = \frac{\sum (x_i - \bar{x})^3}{n(sd(x)^3)}$$

---

[1]The **skewness** function in both of these R packages provides the value for the sample skewness, not the estimate of the population skewness. In your introductory statistics textbooks there may have been some discussion of the difference between these with respect to the standard deviation. A better estimate for the population skewness of a variable **x** is, in R code:

```
n <- length(x)
popskew <- sqrt(n*(n-1))/(n-2)*skewness(x)
```

In standard equations, this is:

$$est.\ pop.\ skew = \frac{\sqrt{n(n-1)}}{(n-2)} sample\ skew$$

As with the standard deviation, the difference is usually only slight, so people tend not to worry about this. For the variable **LENGTH** the population estimate is 1.196.

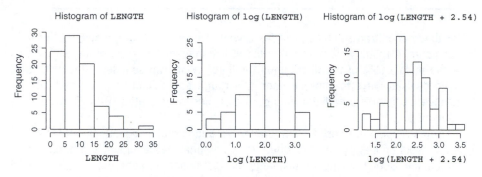

Figure 1.5   The histograms for the variable **LENGTH**, in the left panel, the transformed variable **log(LENGTH)**, in the middle panel, and transformed variable **log(LENGTH + 2.54)**, in the right panel

We will use **e1071**:

```
install.packages("e1071")
library(e1071)
skewness(LENGTH)
```

```
[1] 1.174819
```

which is a pretty high skew (anything over 1 is usually considered high). An equation for standard error of skewness assuming the data are normally distributed is:

$$se_{skewness} = \sqrt{\frac{6(n-2)}{(n+1)(n+3)}}$$

This produces 0.256 for these data. Sometimes an approximation, $\sqrt{6/n}$, is given, and this yields 0.266, so very similar. This means an approximate 95% confidence interval for these data is: $1.17 \pm 2(.26) = 1.17 \pm .52 = (0.65, 1.69)$. One of the problems with this approach is the assumption of normality when it is likely that you are interested in distributions which are not normal.

   A *bootstrap sample* is one taken from the observed sample where you randomly choose one item, record its value, and then return this item to the sample. You then randomly choose a second item, record, and return, and repeat this until you have a sample as large as the original sample. Some items will be chosen multiple times because each time an item is randomly chosen. You can then calculate whatever statistic you want for this bootstrap sample (i.e., the mean, skewness, the *F* from an ANOVA).

   Using a computer you can repeat this procedure thousands of times and record the relevant statistics for each bootstrap sample. The distribution of the statistics

(Cont'd)

for these bootstrap samples provides a way of measuring the precision of your estimates. A rough way to estimate the 95% confidence interval of any statistic is the middle 95% of the distribution for these bootstrap samples. Statisticians have devised ways to improve upon this rough approximation to estimate the confidence intervals and the bias-corrected and accelerated (BCa) method is used here.

Bootstrapping is a very flexible procedure and is rapidly gaining in popularity. It can work on many problems where the traditional mathematical approach has difficulties. Bootstrapping supplements the mathematical approach to statistics with the brut computing force of being able to create thousands of bootstrap samples in seconds!

In these situations a bootstrap estimate can be useful (for a 'leisurely' introduction see Efron & Gong, 1983). We discuss bootstrapping more throughout this book and in the box above. Bootstrapping runs the test over and over by resampling values from those observed data (with replacement). A distribution of the relevant statistic, here skewness, is created. The middle 95% of them is a rough estimate of the 95% confidence interval. More advanced techniques create what are generally thought of as better intervals, and at the time of writing the bias-corrected and accelerated, or BCa, intervals are preferred. We use the package **boot** (Canty & Ripley, 2008). This package is part of CRAN and is automatically installed with R. The **boot** function requires the following: first the variable or variables that you will be using, then the function which you wish to apply to that variable, and then the number of bootstrap samples (here **R=1000**). The complex part is writing the function. For skewness you need to say **function(x,i) skewness(x[i])**. The **x** stands for the variable **LENGTH** and the **i** is for the 1000 replications. When there are multiple variables and more complex functions this can become more difficult, but we will address this in later chapters. Running the function **boot** creates a bootstrap object that can be placed within other functions.

One of these functions, **boot.ci**, produces the most commonly used confidence limits. Differences between these intervals are described in DiCiccio and Efron (1996) and other reviews of bootstrapping. We just concentrate on the BCa estimates. If you had just wanted to print these you could have typed: **boot.ci(lengthboot, type="bca")**.

```
library(boot)
lengthboot <- boot(LENGTH,function(x,i) skewness(x[i]), R=1000)
boot.ci(lengthboot)


BOOTSTRAP CONFIDENCE INTERVAL CALCULATIONS
Based on 1000 bootstrap replicates

CALL :
boot.ci(boot.out = lengthboot)
```

```
Intervals :
Level          Normal                 Basic
95%       ( 0.644, 1.826 )       ( 0.627, 1.765 )
Level          Percentile             BCa
95%       ( 0.584, 1.723 )       ( 0.714, 1.962 )
Calculations and Intervals on Original Scale
Some BCa intervals may be unstable
```

Whichever estimates are used, this variable is skewed (skewness is zero for perfectly symmetric data). Because of this skewness we might want to transform the data; the natural logarithm (**log**) is a popular choice for positively skewed data. The middle panel of Figure 1.5 shows the distribution for this transformed variable is more symmetrical than the untransformed variables (in the left panel), but now its tail seems drawn out to the left. It appears negatively skewed and the statistics confirm this.

```
lnlength <- log(LENGTH)
skewness(lnlength)
```

```
[1] -0.4703981
```

The 95% confidence interval just overlaps with zero if using the traditional ±2se approach and intervals that do not overlap with zero with the bootstrap method. Because bootstrap estimates depend on the particular bootstrap samples chosen, if you run a bootstrap confidence interval several times you may find some of them do overlap (but most will not). If we add an inch to each chile (or 2.54 cm) and then log the data, the skewness is near zero (**skewness(log(LENGTH+2.54))** produces 0.04). This type of transformation is described in the classic regression book by Mosteller and Tukey (1977), where the 2.54 is called a starting value. It is also referred to as a flattening constant, delta, and a Bayesian flat prior, depending on the situation in which it is used. Figure 1.5 shows the three histograms next to each other. To tell the computer to print multiple graphs on the same page you have to use the **par(mfrow=c(1,3))** command. This is a tricky command to remember, but it is often used so is worth memorizing. This tells R that the graph window should have 1 row of graphs and 3 columns of graphs. If you had wanted the graphs in a 2 x 2 grid, you would have typed: **par(mfrow=c(2,2))**. When you are done with any of these multiple-graph-figures it is worth returning this to its default **par(mfrow=c(1,1))**, which has just one graph per figure. You will need to re-shape the graph window to make the figures look like those that are printed in this book. If you right-click on the window you can copy it as a metafile or a bitmap and paste it into other documents in packages including Word and PowerPoint. The following code makes Figure 1.5.

```
par(mfrow=c(1,3))
hist(LENGTH)
hist(log(LENGTH))
hist(log(LENGTH+2.54))
par(mfrow=c(1,1))
```

To end a session, you should detach the data set, here **detach(chile)**. To quit, type **q()**. This is one of the few R functions that has no input into it.

This was an incredibly brief introduction to a very powerful statistics environment. As shown by the rapidly increasing list of R books on http://www.r-project.org/doc/bib/ R-books.html, this free and flexible environment is growing in popularity.

SOME WORDS/CONCEPTS WORTH REMEMBERING

R concepts

- functions: the verbs of R that are applied to objects;
- objects: everything in R;
- CRAN: Comprehensive R Archive Network;
- mirrors: where to download R stuff from;
- packages: sets of functions written for users.

R functions

- **read.spss** and **read.table**: to read SPSS and text files;
- **install.packages**: to access packages from CRAN;
- **library**: to make packages active;
- **seq** and **:**: to make sequences of numbers;
- **par(mfrow=c(7,4))**: to print with 7 rows and 4 columns of graphs;
- **log**: the logarithm function;
- **<-**: to assign objects to each other;
- **c**: to concatenate (or combine) objects.

Statistical concepts

- skewness: a measure of a distribution's symmetry;
- bootstrap: a modern method for estimating the precision of almost anything.

FURTHER READING

Crawley, M. J. (2005). *Statistics: An introduction using R*. Chichester, UK: Wiley. This is a great introduction. Although not written for psychologists, it is still excellent and very clear. Professor Crawley is actually a plant ecologist looking at the interactions between plants and animals, and has books on things like the 'Flora of Berkshire'. http://www3.imperial.ac.uk/naturalsciences/ research/statisticsusingr

Crawley, M. J. (2007). *The R book*. Chichester, UK: Wiley. 950 pages of R and it's legal! This is part reference part teaching book. It is more advanced than his 2005 book, not in terms of statistical knowledge, but in terms of computing skills. http://www.bio.ic.ac.uk/research/ mjcraw/therbook/index.htm

Fox, J. & Anderson, R. (2005). Using the R statistical computing environment to teach social science statistics courses. http://socserv.mcmaster.ca/jfox/Teaching-with-R.pdf and see also http://socserv.mcmaster.ca/jfox/Courses/R-course/index.html. John Fox has written much about R and statistics.

Venables, Smith and the R Development Core Team (2008). *An introduction to R*. Free on http://cran.r-project.org/doc/manuals/R-intro.pdf.

An Amazon list on learning R:
http://www.amazon.com/Learn-the-R-statistics-software/lm/244T3243F9I31/ref=cm_lmt_srch_f_2_rsrsrs0/102-3396071-7592139

The R team recommends various packages for different areas. See:
http://cran.r-project.org/src/contrib/Views/
http://cran.r-project.org/src/contrib/Views/SocialSciences.html

The R help facilities do not tell you about the statistical issues, just how to run the functions. Much R information can be downloaded from the R web site.

The code in this book was run using R 2.4–2.7, but we will update the code if it stops working (if someone tells us!). It is best to use the most up-to-date non-beta version if you are a non-expert.

Note: If using a word processor to type commands, and then pasting them into R, be careful that the symbols you type are not being changed. For example, some word processors might change <- to a single character for an arrow, ←, which R will not understand. This facility can be turned off (in Word, from the Tools/AutoCorrect toolbar). Other problem symbols are ", ', and ellipses (three dots). The line breaks are also sometimes not copied correctly. WordPad and Notepad, which have fewer auto-correcting procedures, are often better to use than the more encompassing word processing packages like Word and WordPerfect. Alternatively, you can use text editors designed for R like Tinn-R (http://www.sciviews.org/Tinn-R/ and http://sourceforge.net/projects/tinn-r).

2

The basic regression

Learning objectives

1. Describe the simple linear regression.
2. Show some R code relevant to regression.
3. Create variables with normal and other distributions.
4. Write data to files.
5. Review the topics to be covered in the remainder of the book.

The word 'regression' gets used by different people in different ways. While it is a very general procedure, people are usually first introduced to a simple linear regression of the form:

$$y_i = \beta 0 + \beta 1 x_i + e_i$$

This equation could either represent a linear relationship between two variables measured on different scales or a difference between the means of y_i for two groups (where x_i could be coded as $x_i = 0$ for one group, and $x_i = 1$ for the other group – what is often called dummy variable coding). These two situations are depicted in Figure 2.1 for data created for illustrative purposes.

Because learning how to use R is an important part of this book, we will list the code for making all figures and for all analyses (in **black**). We do this because learning from examples and adapting the code for your own purposes is one of the best ways for learning R. In each chapter we will add extra important information about R, but you should also search yourself using the help facilities within R. To find out what a function means write **help(xxxx)**, replacing **xxxx** with the name of the function. So, **help(mean)** provides help about the function **mean**. The command **?mean** also provides help. If you do not know the name of the function you can search the help facilities with: **help.search("concept")**. You can also use the search facilities on the CRAN web page. The help facilities are written for statisticians so sometimes may seem abstract or too filled with statistical jargon to be easily understood. Books like this fill in some of these gaps.

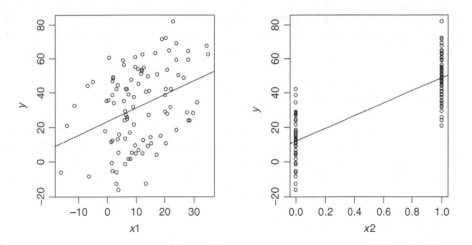

Figure 2.1   Scatterplots depicting simple linear regressions

In this chapter we cover creating new variables from probability distributions and we use the **rnorm** function. So, if you want to know more about this function you would type **help(rnorm)** (or **?rnorm**) and the computer would tell you that **rnorm** creates normally distributed random variables. In the code below the 100 means there are 100 cases, the first 10 is the population mean, and the second 10 the population standard deviation. These are the characteristics of the population from which the variable **x1** is created. If you want to switch the order of the arguments within a function you have to tell R what each means, so **rnorm(10,20,5)** is the same as **rnorm(mean=20,sd=5,n=10)**. **x2** is created using **rbinom(100,1,.5)**. It says to create a binomial variable with 100 cases based on 1 flip of a single coin where the coin has a .5 chance of landing heads (and heads=1, tails = 0). Since a probability cannot be above 1 or below 0, if we had written **rbinom(100,1,1.5)** the computer would have created a variable with 100 missing values (labeled NaN, which stands for 'Not A Number'). We mostly keep with the default settings because that is simpler for teaching purposes and requires less typing (and therefore fewer typos). We have set the random seed so that the data produced will be the same each time you run this code (though the numbers may be different with different versions of R). The third variable is called **y** and it is a combination of **x1** and **x2** with additional random normally distributed error. It also has 100 cases.

```
set.seed = 121
x1 <- rnorm(100,10,10)
x2 <- rbinom(100,1,.5)
y <- x1 + x2*40 + rnorm(100,0,10)
```

Next we can run some simple linear regressions between **y** and the x variables. The basic regression function in R is **lm**. **lm** stands for linear model. **lm(y~x1)** says to run the regression $y_i = \beta 0 + \beta 1 x 1_i + e_i$. There are more advanced regressions that will be introduced during this book which build on this syntax, but they all have the form that the tilde symbol, ~, separates the response variable, here **y**, from the variables that you are using to predict it, here **x1**. This notation is based on one developed by Wilkinson and Rogers (1973). When we run a regression it creates a **lm.object**. Here we run two regressions and store the results in **reg1** and **reg2**. As will be shown, we can use these objects in other functions.

```
reg1 <- lm(y~x1)
reg2 <- lm(y~x2)
```

Next we are going to draw scatterplots of these data. The **par(mfrow=c(1,2))** means place the graphs in a 1 x 2 grid. A scatterplot can be made with the **plot** function which is a versatile function. The type of plot depends on the type of objects to which it is applied. If two numeric variables are entered into **plot**, as is the case here, a scatterplot is drawn.[1] The code below creates Figure 2.1.

```
par(mfrow=c(1,2))
plot(x1,y)
abline(reg1)
plot(x2,y)
abline(reg2)
par(mfrow=c(1,1))
```

The **abline** functions place the regression lines onto the scatterplots. There are a few functions within R that draw lines onto scatterplots. **abline** is for straight lines.

Traditionally, the situation shown in the left panel of Figure 2.1 is called a simple linear regression, and on the right is a *t* test. However, both are the same model – the one depicted by the equation above. There are several assumptions for assessing these models. Many of the topics discussed in this book are about extensions which address these assumptions. These two situations are likely to have been dealt with in your introductory statistics courses, but it is worth repeating them in order to show how R can estimate them.

The default diagnostic checks for the model, $y_i = \beta 0 + \beta 1 x 1_i + e_i$, are shown in Figure 2.2. Figure 2.2 was made with the **plot** function. If a **lm.object** is entered into the **plot** function it assumes that you want these four graphs. Because there are four graphs we have told R to present them in a 2×2 grid.

```
par(mfrow=c(2,2))
plot(reg1)
```

The Q–Q (quantile–quantile) plot in Figure 2.2 shows systematic deviation from normality. If the residuals were normally distributed, we would expect the points in the Q–Q plot to form a straight line near the diagonal with only non-systematic deviation. Systematic deviations can arise for several reasons, but because we have made up the data ourselves we know that the deviations are likely to be due to the model (which includes just **x1**) being incomplete. It should also include **x2**. This suggests that we should try adding the variable **x2**, creating what is usually called a multiple regression: $y_i = \beta 0+$

---

[1]For comparison, if **x2** were treated as a factor the plot command would draw two boxplots. To show this you need to type: **plot(as.factor(x2),y)**.

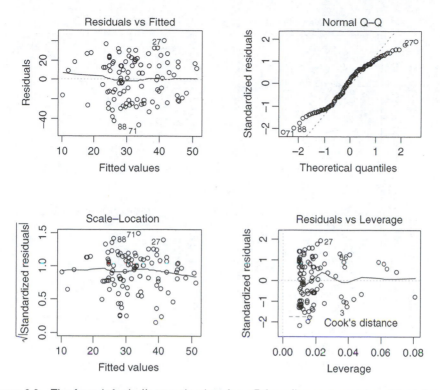

Figure 2.2   The four default diagnostic plots from R for a linear regression object, for the model $y_i = \beta 0 + \beta 1x1_i + e_i$

$\beta 1x1_i + \beta 2x2_i + e_i$ which in R we write as **lm(y~x1+x2)**. When we examine the diagnostic graphs for this, there appears no systematic deviation in the Q–Q plot (see Figure 2.3).

```
reg3 <- lm(y~x1+x2)
plot(reg3)
```

Researchers are usually interested in the parameter estimates for these models. In the code above we have created three regression or **lm** objects: **reg1,reg2**, and **reg3**. The way R works is to use functions that take information from these objects and, in the case of **plot**, creates the plots above, or in the case of **summary** provides the parameter estimates (the coefficients), the multiple correlation coefficient, and the like. So, **summary(reg3)** produces:

```
Call:
lm(formula = y ~ x1 + x2)

Residuals:
      Min        1Q     Median        3Q       Max
 -26.5775   -6.7035   -0.8574    7.7245   23.5401
```

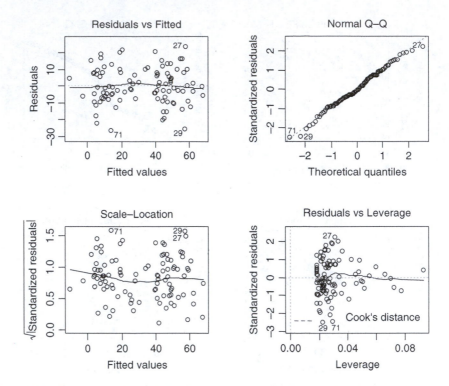

Figure 2.3 The diagnostic plots for the regressions including both x1 and x2

```
Coefficients:
              Estimate Std. Error t value Pr(>|t|)
(Intercept)    2.6183     1.9887    1.317    0.191
x1             0.8147     0.1047    7.785 7.68e-12 ***
x2            36.9657     2.1490   17.201  < 2e-16 ***
---
Signif. codes:  0 '***' 0.001 '**' 0.01 '*' 0.05 '.' 0.1 ' ' 1

Residual standard error: 10.69 on 97 degrees of freedom
Multiple R-Squared: 0.7852,     Adjusted R-squared: 0.7807
F-statistic: 177.2 on 2 and 97 DF,   p-value: < 2.2e-16
```

> **Note:** The `7.68e-12` means 7.86 x 10^{-12} or 0.00000000000786. This notation is useful for very small numbers, like this, or very big numbers. Most of the time you encounter this notation it will be for *p* values very near to zero so you can simply write '*p* <.001'.

These estimates are all pretty close to those used to create the data (i.e., the population parameters). There are a few different equations for the adjusted R^2. Here it is:

$$R^2_{adj} = 1 - \frac{(1 - R^2)(n - 1)}{n - k - 1}$$

where k is the number of predictor variables. The population values are: $y_i = 0 + 10\,x1_i + 40\,x2_i + e_i$. If you want to purposefully exclude the constant you have to write: **reg4 <- lm(y~x1+x2-1)**. If you want to add the interaction term (which here means that the slope for **x1** is different for the two groups), it is **reg5 <- lm(y~x1+x2+x1:x2)** or **reg5 <- lm(y~x1*x2)**. The **x1:x2** means the interaction between **x1** and **x2**, and **x1*x2** means the interaction of **x1** and **x2** and all the effects nested within this (so **x1*x2*x3** is the same as **x1:x2:x3+x1:x2+x1:x3+x2:x3+ x1+x2+x3**).

You can see if a nested model fits significantly better than another with the **anova** function.[2] The one with no interaction nor any intercept is compared with the regression with these by: **anova(reg4,reg5)**:

```
Analysis of Variance Table

Model 1: y ~ x1 + x2 - 1
Model 2: y ~ x1 + x2 + x1:x2

     Res.Df        RSS    Df Sum of Sq         F    Pr(>F)
1        98    11285.0
2        96    11033.1     2      251.8    1.0956    0.3385
```

The second model accounts for more of the variation (RSS is residual sum of squares which drops by 250), but it uses 2 more degrees of freedom (for the intercept and the interaction). This difference is nonsignificant. The difference divided by the residual sum of squares of the first model is the popular effect size, partial eta-squared. Here it is: $251.9/11285.0 = .02$. We could write:

Including the interaction and intercept did not significantly improve the fit of the model, $F(2, 96) = 1.10, p = .34, \eta_p^2 = .02$.

The **anova** command (remember that R is case sensitive, **Anova(reg4,reg5)** will produce an error) can be used with non-nested models, like **reg1** and **reg4**, but care should be taken when comparing these. This is a big issue in, for example, the structural equation model literature, where there is discussion comparing things like BIC and AIC. BIC is the 'Bayesian Information Criterion'. AIC is 'An Information Criterion', developed by Akaike, so usually called Akaike's Information Criterion. These are revisited in more detail in Chapter 5 on model selection.

It is worth mentioning a common misunderstanding about what a 'linear model' is, and this relates to the interaction model. The interaction is constructed by multiplying the two x variables together. We could have also included $x1_i^2$ (in R **x1^2**) or even $\sin(x1_i)$ (in R **sin(x1)**) or any function of **x1**, and it would have still been a 'linear model.' The word 'linear' is referring to the β values, not the x variables. It is often useful to

---

[2]The phrase analysis-of-variance (ANOVA) refers to comparing the ratio of two variances. This is the final step in the popular statistical procedure for comparing means, which is often also called ANOVA. The ANOVA procedure for comparing means is run in R with a function called **aov** or with **lm** (see Chapter 3).

re-write regression equation, $y_i = \beta 0 + \beta 1 x 1_i + \beta 2 x 2_i + \beta 3\, x 1_i x 2_i + e_i$, in longhand matrix form as:[3]

$$[y_1 \;\; y_2 \; ... \; y_n] = [\,\beta 0\; \beta 1\; \beta 2\; \beta 3\,] \begin{bmatrix} 1 & 1 & ... & 1 \\ x1_1 & x1_2 & ... & x1_n \\ x2_1 & x2_2 & ... & x2_n \\ x1x2_1 & x1x2_2 & ... & x1x2_n \end{bmatrix} + [\,e_1 \;\; e_2 \; ... \; e_n\,]$$

or in shorthand as $\mathbf{Y} = \boldsymbol{\beta}\mathbf{X} + \boldsymbol{e}$, where the **bold** characters represent matrices. The 'linear' refers to being able to separate the β values like this. So, $y_i = \beta 0 + x 1_i{}^{\beta 1} + e_i$, looks relatively simple, but is non-linear. The traditional least square methods do not work for most non-linear models like this. All the main statistical packages (R, SPSS, etc.) have procedures for non-linear regression. Most non-linear regression procedures work by beginning with some starting values for the β values, assessing how well the data fit this model, and then tweaking the β values so that the fit is better. These new values are then used and the computer assesses how well the data fit a model with them. This is continued until tweaking the β values no longer improves the fit very much. This is called an iterative procedure, which can be computationally demanding. With modern computers, this computational demand is not a serious drawback, but there are other technical issues and conceptual difficulties. In Chapter 6 we discuss how some specific types of non-linear models can be changed into linear ones and in Chapter 7 how linear models can be made much more flexible. However, other non-linear models are not often used in psychology because most of the theories are so imprecise that nothing is gained by testing a non-linear model, particularly compared with the types of flexible linear models that are introduced throughout this book. This is not a criticism of psychology. As we will see, the extensions to the basic linear models that are covered in this book will take us fairly far. It is just that the way theory progression is done in psychology (exceptions, particularly in psychophysics, exist) does not require these more complex models, and model simplicity is good for science.

Example 2 – Dissociation and suggestibility

- Data: below, from Wright and Livingston-Raper (2001);
- Research question: Are dissociation and memory suggestibility related?;
- Purpose: To illustrate a simple linear regression with R and to create a simple scatterplot.

These data examine whether dissociation (what is high in what used to be called *multiple personality disorder*) predicts suggestibility on an eyewitness memory study. Most of the data in this book are ours. This is because we have these data files handy and are allowed to disseminate them without copyright issues.[4]

Data can be read into R in several ways. In Chapter 1 you were shown how to access them from other files but for small data sets it is often convenient to type the numbers directly into R. Here we assign the 50 values for each variable to **suggest**

---

[3]If you have not used matrix notation before, do not worry. This is the only place in the book where we use it like this. It is useful shorthand, but is not necessary for conceptual understanding.

[4]The website that accompanies this book includes other examples. We encourage you to include your own examples on these pages. Directions for how this should be done are on the website.

and **DES** (DES stands for dissociative experiences scale). The function **c** is often used in R; it stands for concatenation and is used to tell the computer that these numbers form a set. Notice that with long commands they will stretch over multiple lines. In many R sources authors place a **+** at the start of a new line if it is a continuation of a previous command. We do not. Where it is unclear we indent the continuation lines.

```
suggest <- c(12, 2, -2, 5, 10, -13, 10, -3, 4, 9, 13, 6, 18, 12, 14,
    -6, 0, 5, 6, 19, -5, 8, 5, 14, -1, -2, 10, 17, 2, 10, -1, 21,
    14, 4, 20, 24, 5, -12, 8, 7, 0, 2, 7, -1, 12, 4, 0, 19, 8, -12)
DES <- c(45.00,49.28,38.57,55.71,48.21,14.60,12.86,43.50,27.10,
    46.40,46.40,35.00,55.00,52.85,46.40,18.21,36.79,48.93,45.50,
    22.10,31.43,32.50,52.14,18.21,22.50,27.14,16.67,14.64,39.64,
    16.78,34.28,69.64,38.90,23.20,41.67,62.86,47.00,22.86,31.42,
    48.21,18.57,48.21,40.00,35.00,57.85,50.35,41.07,45.00,48.00,49.28)
```

The values for 50 participants on these two variables are in the same order so that the fourth participant's can be accessed with **suggest[4]** and **DES[4]**.

Next, the simple linear regression is run with the **lm** function, a **lm.object** is stored in **reg**, and **summary** is used to print information about this object. The **lm** function assumes the variables are in the same order, so that the fourth number in **suggest** corresponds to the same person as the fourth number in **DES**. When a **lm.object** is entered into the **summary** function, R produces the most commonly reported statistics for a regression. R prints a * to denote $p < .05$, ** for $p < .01$, etc. Because many people find this a 'poor scientific strategy' (Meehl, 1978: 817) you may want to turn this facility off with **options(show.signif.stars=FALSE)**. The words **TRUE** and **FALSE** can be replaced by **T** and **F** in most R functions, though it is recommended that you use **TRUE** and **FALSE** to avoid confusion. Though in order that commands fit on as few lines as possible we sometimes do not.

```
reg <- lm(suggest~DES)
summary(reg)


Call:
lm(formula = suggest ~ DES)

Residuals:
     Min        1Q    Median        3Q       Max
-20.26046  -5.99737  -0.02292   6.03891  15.92422

Coefficients:
            Estimate Std. Error t value Pr(>|t|)
(Intercept)  -1.1399     3.4105  -0.334   0.7397
DES           0.1908     0.0838   2.276   0.0273 *
---
Signif. codes:  0 '***' 0.001 '**' 0.01 '*' 0.05 '.' 0.1 ' ' 1

Residual standard error: 8.21 on 48 degrees of freedom
Multiple R-Squared: 0.09744,    Adjusted R-squared: 0.07863
```

```
F-statistic: 5.182 on 1 and 48 DF,  p-value: 0.02732
```

Because the coefficient for **DES** (0.1908) is positive, we can tell that in the sample there is a positive relationship between the dissociation scores and memory suggestibility, with a multiple R^2 of 0.09744. To find Pearson's r we can take the square root of this: **sqrt(.09744)** yields 0.3121538. There are several ways to report this result in a manuscript including $t(48) = 2.28$, $p = .03$, and $r = .31$, $n = 50$, $p = .03$. These are reporting the same information (being sensible with the sign, if t is negative r should also be negative) because:

$$r = \sqrt{\frac{t^2}{t^2 + df}}$$

To interpret a regression properly it is necessary to graph the data. Over the last 30 years there has been a large increase in the awareness that graphs are important for statistics. There has also been a large increase in what computers can do to make good graphs (and also in how they can create bad graphs, Wainer, 1984). Figure 2.4 is just the basic scatterplot between these variables. We have added our own labels for the x and y axes. The regression line is plotted using the **abline** function. It is often useful to add text to graphs. The **text** function does this. It requires three arguments: the location on the x axis and on the y axis and the text to write. The text needs to be in quotation marks.[5]

```
plot(DES,suggest,xlab="Dissociation score (0-100)",
 ylab="Suggestibility (-25-25)")
abline(reg)
text(65,-12,"r = .31")
```

After completing your analyses you may want to save the data. At present, there are just the two variables (**DES** and **suggest**) floating around in your current work space. The only way that R knew they were related was that in the **lm** and **plot** functions they were of the same length and placed within these functions together. To create a data set of these two variables we can use the column combine function **cbind**:

```
dataset1 <- cbind(DES,suggest)
```

We could then write these data to our hard drive:

```
write.table(dataset1,"c:\\temp\\dataset1.dat")
```

The \\ are needed when either saving or accessing information from other locations.

The data can be written in the appropriate format for other packages too. We can use the package **foreign** (R Core members et al., 2008), which we used in Chapter 1 to read data from SPSS, to write the data to other packages. If you are using the same computer as you were for Chapter 1, and if the computer hard drive is not cleaned (which happens

---

[5]You can write **text(locator(1),"r =.31")** and then click onto the plot where you want the text. We do not recommend this because if you want to re-make the graph it is nice knowing the exact location of the text.

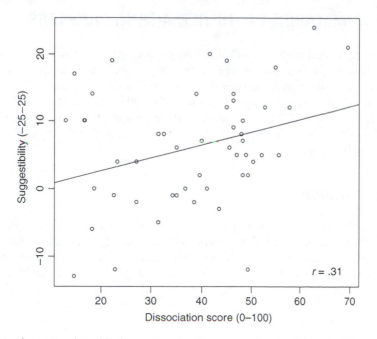

Figure 2.4   A scatterplot with the regression line shown for the Wright and Livingston-Raper (2001) data

on many public access machines), you should not need to re-install this package. But, if you are on a new computer, to access this package type:

```
install.packages("foreign")
```

and it will prompt you to choose a mirror site (see Figure 1.2). Once it is installed to activate all the functions within the package type:

```
library("foreign")
```

To save these data as an SPSS .sav file write:

```
write.foreign(as.data.frame(dataset1),"c:\\temp\\dataspss.dat",
"c:\\temp\\dataspss.sps",package="SPSS")
```

This is a fairly long command. The **as.data.frame(dataset1)** part tells the computer what data to save and the function needs the data to be a particular type of object called a dataframe, so the data are read as this. The second argument tells the computer where to put the data. The third tells it where to write a SPSS syntax file. The fourth argument tells R that this is a SPSS conversion (it also can convert data into SAS and Stata). Within SPSS you can run the syntax file to create the SPSS data file (for information on SPSS, see Field, 2009). Some characters, variable types, missing values, etc., are not copied exactly. As with translating between any types of files care is needed.

BRIEF SUMMARIES OF THE REMAINING CHAPTERS

The basic regression can be extended in many ways, and only a handful of these—the ones we believe are most useful—are covered in this book. We already mentioned that we will not describe non-linear regression. Nor do we discuss multivariate regression (where there are multiple response variables[6]), although the multilevel modeling procedure described in Chapter 8 can be used for multiple response variables. We also do not consider latent variable models. Here we briefly describe the topics that we have included in this book.

Chapter 3. ANOVA as regression

ANOVA (ANalysis Of VAriance) is one of the most common techniques used in psychology. ANOVA is a particular way of conceptualizing a regression. All ANOVAs are regressions and in this chapter the R regression procedures are used to run ANOVAs. A secondary aim of the chapter is to stress that you should not arbitrarily turn a scale variable into a binary variable by dichotomizing. We examine data based on Festinger and Carlsmith's (1959) classic cognitive dissonance study as an example.

Chapter 4. ANCOVA: Lord's paradox and mediation analysis

ANCOVA (ANalysis of COVAriance) can be conceived of either as an extension to ANOVA or as a type of multiple regression. It is used when researchers wish to compare groups on some variable controlling for the level of another variable (the covariate). We examine this by looking at two statistical problems often faced: Lord's paradox and mediation. A recent example from child eyewitness testimony (London et al., in press) is used. We also discuss writing functions in R, which is an important part of using R.

Chapter 5. Model selection and shrinkage

Sometimes you have several predictor variables and wish to select a subset of these for reasons both of parsimony and that simpler models are often more reliable than more complex ones. Methodologists are uniformly critical of the automatic stepwise procedures that are used in many of the popular packages (like SPSS). There are several alternative techniques. The example data come from Ayers et al.'s (2007) study of PTSD after child birth in mothers and fathers.

Chapter 6. Generalized Linear Models (GLMs)

There is a class of techniques which are often appropriate when your response variable is binary, a count, or a proportion. These can all be estimated in a similar and efficient way. The most common of these (other than the basic regression which is a special case of it)

---

[6]We refer to variables on the left side of the equal sign (or ~ in R functions) as response variables and those on the right side as predictor variables and covariates. Sometimes these are referred to as the DV and the IVs, for dependent and independent variables.

is the logistic regression which is appropriate when the response variable is either binary or a proportion. We briefly examine the most common types of GLMs, and then look in more detail at an example from juror decision making which uses logistic regression to estimate the meaning of 'reasonable doubt' (Wright & Hall, 2007).

Chapter 7. Regression splines and Generalized Additive Models (GAMs)

Additive models involve piecing together different curves, which are called *splines*, to map out the relationship between each predictor variable and the response variable. These are very flexible models and are most valuable either for graphically exploring data or when not wanting to make many assumptions about one predictor variable when looking at more specific models with another predictor variable. We use two examples, one exploring the gender gap in income (Berndt, 1991) and the other looking at ways of detecting deception.

Chapter 8. Multilevel modeling

Often psychology data have a multilevel structure. The prototypical example is testing children who are nested within their school, so are not independent of their classmates. But, there are numerous other situations where multilevel modeling is appropriate. We look at children's exercise in several schools in the Sussex (UK) area (Hill et al., 2007) and the relationship between response time and accuracy in cross-race memory research (Wright et al., 2003).

Chapter 9. Robust regression

It has long been known that the traditional ANOVA and regression are not robust, meaning that they are highly affected by outliers. There are many robust alternatives that are less affected by outliers, and tend also to be more powerful. We begin by using Spearman's ρ on crime data in the Sussex area. Wilcox's skipped correlation is then shown, with UN data on children's well being, as an example method which removes bivariate outliers. Finally, robust regressions, which lessen the impact of large residuals, are used to analyze Anscombe's (1973) classic data.

SOME WORDS/CONCEPTS WORTH REMEMBERING

R concepts

* creating variables: assigning values to a variable name.

R functions

* `help`: to learn about functions;
* `write.table`: writes data to files;

(Cont'd)

- **lm**: runs basic regressions;
- **plot**: makes lots of different kinds of figures;
- **anova**: compares two or more regression objects;
- **summary**: provides output from statistical procedures;
- **rnorm** and **rbinom**: creates normal and binomial variables;
- **abline**: draws straight lines on scatterplots;
- **:** and *****: for interaction terms in regressions;
- **text**: adds text to plots.

Statistical concepts

- linear regression: a model linear in the parameters (the β values).
- Q–Q plot: a plot to check the normality of residuals.

FURTHER READING

Fox, J. (2002). *An R and S-Plus companion to applied regression*. Thousand Oaks, CA: Sage Publishing. This is the book most similar to ours as far as content. This is a very good book. It assumes more statistical knowledge than we do.

Mosteller, F. & Tukey, J. W. (1977). *Data analysis and regression: A second course in statistics*. Reading, MA: Addison-Wesley Publishing Company. This book is a classic textbook. It is more advanced than Wright and London (2009).

Wright, D. B. & London, K. (2009). *First (and second) steps in statistics* (2nd Ed.). London: Sage Publications. Most introductory textbooks cover basic regression. Introductory textbooks vary in how much statistical knowledge is assumed. This one is on the beginner side of this scale.

3

ANOVA as regression

Learning outcomes

1. Running a oneway ANOVA as a regression;
2. Using different contrasts in R;
3. Learning why you should not categorize variables unless necessary;
4. Learning how R treats factors and numeric variables.

This chapter covers two almost opposite topics. This first is what to do if you have predictor variables that are categorical. This involves using regression like an ANOVA. The second is that unless the data really are just categorical you should usually not treat them as categorical. We use data from two studies. The first example is based on the classic cognitive dissonance study by Festinger and Carlsmith (1959) showing that giving people insufficient reward for lying can create dissonance. The second example is from a recent study by London et al. (in press, Exp. 2). The primary purpose of their research was to look at the long-term effects of a suggestive interview. Here just the correct free recall utterances are examined.

Example 3 – cognitive dissonance

- Data: From Wright and London (2009) and constructed to be similar to Festinger and Carlsmith (1959). Some of the description closely follows Wright and London.
- Research question: Does giving a small monetary reward create cognitive dissonance for doing an unpleasant task?
- Purpose: To show how to run ANOVAs with the `lm` function and how to use contrasts.

In one of the classic studies of social psychology, Festinger and Carlsmith (1959) had participants spend about one hour putting spools onto a tray and turning square pegs a quarter rotation. It was designed to be boring and was! After participants finished this tedious task, the experimenter pretended as if the study was over and gave them a pretend debriefing. Participants were told that there were two groups in the study and that they were in the control group who had received no information before the study. They were told that people in the other group spoke with a confederate (someone working for the experimenter but pretending to have just taken part in the study as a participant) who told them that the experiment was enjoyable. At this point, the real study was just beginning.

There were three groups in Festinger and Carlsmith's study. There was a control group who after the 'debriefing' were ushered into a waiting room. There were two experimental groups. For each of these the experimenter explained that the usually reliable confederate had phoned saying that he could not make it. The experimenter asked if the participant would help out and tell a female participant who was waiting in the next room that the experiment was enjoyable. One group was paid \$1 and the other was paid \$20.[1] Most complied, although a few said they were suspicious and their data were discarded.[2] After the participant either waited in an empty room (control group) or told the confederate that the boring task was enjoyable, they thought they were done. On leaving the building, they were informed that the department monitors all experiments and asked if they would fill out a questionnaire. This in fact was an integral part of the study. It included a question asking them, on a -5 to $+5$ scale, how interesting and enjoyable the study (the boring spools and pegs tasks) was. The prediction from Festinger's cognitive dissonance theory is that those paid only \$1 were more likely to say the task was enjoyable compared with the other groups.

Festinger and Carlsmith had 20 people in each condition. We have recreated these data so they closely resemble their original data.

```
control <- c(0,-3,3,2,-2,-1,2,3,-3,-5,2,-3,3,0,-2,
   -2,-2,-2,-1,2)
dollar1 <- c(3,1,1,3,2,3,3,2,2,2,2,2,-4,4,0,-3,
   4,1,1,-2)
dollar20 <- c(1,2,3,0,1,3,0,-2,2,1,0,0,-1,-2,-1,-4,
   -3,-1,0,0)
```

These need to be combined into a single response variable:

```
enjoy <- c(control,dollar1,dollar20)
```

We need to create a variable for condition where the first 20 values are for the control group, the next 20 are for the \$1 group, and the final 20 are \$20 group. The **rep** function does this. The **each=20** means to write each one of these 20 times.

```
group <-
   as.factor(rep(c("control","$1","$20"),each=20))
```

We have to tell R that we want it to treat this sequence as a categorical variable, which in R terminology is a **factor**.

---

[1]All were asked for this money back at the end of the real experiment. Festinger and Carlsmith (1959: 207) said all 'were quite willing' to do this. Would this replicate?

[2]Another participant's data were discarded because he asked for the female's phone number and said 'he would call her and explain things' (p. 207) and wanted to stay around until she was done with the experiment so they could talk, presumably wanting to 'debrief' the young lady (who was the actual confederate). After the actual experiment, participants were debriefed with the female confederate present. The participant slept alone that night.

Several R functions can be used to explore these data. The first might be to look at the means and standard deviations. The **tapply** function can be used to look at the value of any statistic for a variable broken down by groups. Here, the **mean** and the **sd** (standard deviation) of **enjoy** are printed for each value of **group**.

```
tapply(enjoy,group,mean)
```

```
$1        $20         control
1.35      -0.05       -0.45
```

```
tapply(enjoy,group,sd)
```

```
$1          $20          control
2.158825    1.848897     2.438183
```

Because **group** is a factor the **plot** function makes separate boxplots for each group (Figure 3.1). We have added labels to the *y* and *x* axes. The **cex.lab=1.3** makes these labels 1.3 times larger than their default.

```
plot(group,enjoy,ylab="Enjoyment (-5 to 5)",
    xlab="Condition",cex.lab=1.3)
```

Next, the ANOVA can be run. In R there is an ANOVA function, **aov**.

```
anova1 <- aov(enjoy~group)
summary(anova1)
```

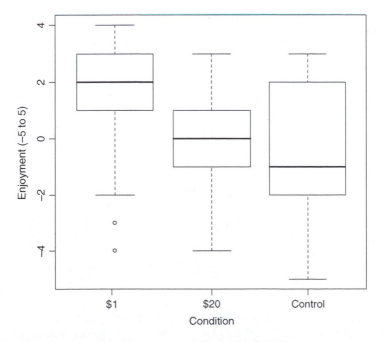

Figure 3.1   Boxplots based on Festinger and Carlsmith (1959)

```
            Df    Sum Sq   Mean Sq  F value  Pr(>F)
group        2    35.733   17.867   3.8221   0.02769   *
Residuals   57   266.450   4.675
---
Signif. codes:  0 '***' 0.001 '**' 0.01 '*' 0.05 '.'
     0.1 ' ' 1
```

This is the standard ANOVA table that you will have been taught to make during introductory statistics courses. You will probably have also been taught to calculate the proportion of the total Sums of Squares (SS) accounted for by the model. The total SS is 35.733 + 266.450 = 302.183. The ratio is 35.733/302.183 = .118. In ANOVA terminology this is called η^2 (eta squared) and it is a common measure of effect size (and is related to the partial eta squared). Another effect size sometimes reported in these situations is ω^2 (omega squared). It is like η^2 but adjusted to take into account the complexity of the model. An equation for this is:

$$\omega^2 = \frac{SS_b - (k-1)MSe}{SS_{total} + MSe}$$

From the data above, this would be: **(35.733 - (3-1)*(4.675))/(35.733+ 266.450+4.675)** $= 0.08597788$ or 8.6%. ω^2 will always be smaller than η^2, in the same way as adjusted R^2 is always smaller than R^2. The values of adjusted R^2 and ω^2 are different, but they have a similar meaning.

The same model can be run with the **lm** function.

```
anova2 <- lm(enjoy~group)
summary(anova2)

Call:
lm(formula = enjoy ~ group)

Residuals:
    Min      1Q    Median     3Q      Max
   -5.35   -1.55    0.05      1.65    3.45

Coefficients:
               Estimate Std. Error t value  Pr(>|t|)
(Intercept)     1.3500   0.4835      2.792   0.00711   **
group$20       -1.4000   0.6837     -2.048   0.04521   *
groupcontrol   -1.8000   0.6837     -2.633   0.01088   *
---
Signif. codes:  0 '***' 0.001 '**' 0.01 '*' 0.05 '.'
     0.1 ' ' 1

Residual standard error: 2.162 on 57 degrees of freedom
Multiple R-Squared: 0.1183, Adjusted R-squared: 0.08731
F-statistic: 3.822 on 2 and 57 DF, p-value: 0.02769
```

This output looks very different from the ANOVA output, but a couple of things look the same. For example, the p value at the end is the same; the multiple R^2 is the same as what we calculated for η^2, and the degrees of freedom for both the model and the residuals are the same. In fact, the model is the same, just the output looks different. The **anova** function produces the ANOVA table from information stored in an **lm.object**.

anova(anova2)

```
Analysis of Variance Table

Response: enjoy
           Df    Sum Sq   Mean Sq  F value  Pr(>F)
group      2     35.733   17.867   3.8221   0.02769  *
Residuals  57    266.450  4.675

---
Signif. codes:  0 '***' 0.001 '**' 0.01 '*' 0.05 '.'
     0.1 ' ' 1
```

This shows that ANOVA is just a way of describing the results from a regression. The regression model being evaluated is:

$$enjoy_i = \beta0 + \beta1 \ group\$20_i + \beta2 \ groupcontrol_i + e_i$$

When the R functions **aov** and **lm** encounter a factor like **group**, it creates dummy variables. These are variables with the value 1 for one of the groups and 0 for all other cases. When there are k categories, R creates $k - 1$ dummy variables. Here there are 3 categories so 2 dummy variables are created. The first has the value 1 for the $20 group and 0 for everyone else. The second has the value 1 for the control group and 0 for everyone else. The $1 group has the value 0 for both of these. It is called the *reference* category. R has chosen it as the reference category simply because it is first in its list. These are called *contrasts* and these are a major focus in books on ANOVA. To find out how R plans to create dummy variables for a factor use the **contrasts** function.

contrasts(group)

```
           $20    control
$1         0      0
$20        1      0
control    0      1
```

The mean for the reference category will be the estimate for the intercept: 1.35. The mean for the $20 group will be the intercept plus the estimate for this coefficient $(1.35 - 1.40 = -0.05)$. The statistics for this coefficient provide a test for whether this group is different from the reference group. It is, $t(57) = 2.05$, $p = .05$. Similarly, the mean for the control group is $1.35 - 1.80 = -0.45$, $t(57) = 2.63$, $p = .01$. This is slightly different from running t tests comparing groups and this can be seen because the number of degrees of freedom is for the entire sample, not just the two groups.

You may want to change how R constructs contrasts. The **contrasts** function also allows this to be done. Often you want to compare each group to the control group. To do this, you need to construct a matrix like the one shown above but with zeroes for the control row and assign this to the contrasts. The **dim** function says what dimensions the contrast matrix has.

```
wcontrol <- c(1,0,0,0,1,0)
dim(wcontrol) <- c(3,2)
contrasts(group) <- wcontrol
contrasts(group)
```

```
          [,1]      [,2]
$1         1         0
$20        0         1
control    0         0
```

Now when the regression is run the coefficients compare each group with the control group.

```
summary(lm(enjoy~group))

Call:
lm(formula = enjoy ~ group)

Residuals:
    Min      1Q     Median     3Q      Max
   -5.35   -1.55     0.05     1.65     3.45

Coefficients:
                Estimate Std. Error t value Pr(>|t|)
(Intercept)     -0.4500     0.4835   -0.931   0.3559
group1           1.8000     0.6837    2.633   0.0109  *
group2           0.4000     0.6837    0.585   0.5608
---
Signif. codes:  0 '***' 0.001 '**' 0.01 '*' 0.05 '.'
    0.1 ' ' 1

Residual standard error: 2.162 on 57 degrees of freedom
Multiple R-Squared: 0.1183, Adjusted R-squared: 0.08731
F-statistic: 3.822 on 2 and 57 DF, p-value: 0.02769
```

We can see the $1 group is different from the control group, $t(57) = 2.63$, $p = .01$, which we knew from before when the $1 group was the reference category. We now have a test comparing the control group and the $20 group, which is non-significant, $t(57) = 0.59$, $p = .56$. Note that the overall F value and other statistics are all the same. It is worth noting that we have done three tests between the different pairs of groups. When you increase the number of statistics tests you increase the chances of incorrectly rejecting a true null hypothesis (a type 1 error) and failing to reject a false null hypothesis (a type 2 error). Therefore you should be cautious when conducting many of tests. Some books suggest controlling the chance of a type 1 error by requiring a lower p value, but this increases the chances of a type 2 error. This has become a major concern for some modern statistical procedures in, for example, bioinformatics and brain imaging studies, where you may have hundreds or thousands of statistical tests.

R has many built in contrasts that you can choose from. This is a large and sometimes complicated area. It is covered well in most books on ANOVA. It is less important with the typical regressions because usually the predictor variables are not multi-valued categorical variables. Sometimes they are ordinal. If you tell R that a variable is ordinal by **group <- as.ordered(group)** then the default contrasts in R are polynomial contrasts (see Chapter 7).

Example 4 – Children's forgetting 1

* Data: from London et al. (in press, Exp. 2)
* Research question: Do younger or older children forget more rapidly?
* Purpose: To show that categorical predict variables can be used in regression, but that you should not unless they really are categorical. This example is used to show how to switch between different types of data.

Children, aged 5 to 9 years-old, participated in a magic show. Two weeks later they took part in an exit interview where they were asked about the magic show. Then, approximately ten months later, they were re-interviewed. London et al. expected most children to recall less after the delay. They wanted to see whether the delay affected children of different ages in similar ways. This can be looked at in several ways. First we look at it by comparing the final amount recalled by different ages, and then we look at it by the difference in the amount recalled between the two times. In Chapter 4 this is looked at with a different statistical procedure, the ANCOVA, and then in Chapter 7 generalized additive models are used to explore these data.

The data file has just three variables (age in months, final score, and initial score) and is stored in a text file. If it were on your computer, say in c:\temp, then the command

```
lordex <- read.table("c:\\temp\\lordex.txt",header=T)
```

would create an object in R called **lordex** which could then be attached. Alternatively, because these data are on the book's web page they can be directly accessed with:

```
webreg <- "http://www.sagepub.co.uk//wrightandlondon//"
lordex <- read.table(paste(webreg,"lordex.txt",sep=""),
   header=T)
attach(lordex)
```

It is often useful to print the **names** of the variables.

```
names(lordex)
```

```
[1] "AGEMOS" "Final" "Initial"
```

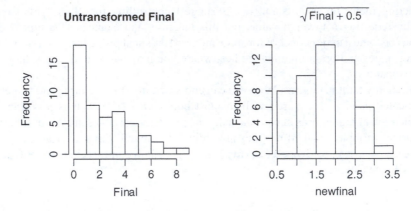

Figure 3.2   Histograms of the untransformed Final and when it is transformed with the square root, plus .5, transformation

We are going to compare the child's age and their final recall so it is worth looking at both of these variables. First we will look at **Final**. The left panel of Figure 3.2 shows a histogram for **Final**. Here is the code for this where we first tell the computer that we want one row with two graphs.

```
par(mfrow=c(1,2))
hist(Final,main="Untransformed Final")
```

It appears positively skewed. Following the procedures used in Chapter 1 we can calculate skewness and the 95% confidence interval for skewness.

```
library(e1071)
library(boot)
skewness(Final)
```

```
[1] 0.6433149
```

```
skewboot <- boot(Final, function(x,i) skewness(x[i]),
    R=1000)
boot.ci(skewboot,type="bca")
```

```
BOOTSTRAP CONFIDENCE INTERVAL CALCULATIONS
Based on 1000 bootstrap replicates

CALL :
boot.ci(boot.out = skewboot, type = "bca")

Intervals :
Level                 BCa
95%         (0.2317, 1.1750)
Calculations and Intervals on Original Scale
```

In Chapters 6 and 7 methods are described that could be used to analyze, directly, the untransformed variable but for the methods described in this and the next chapter it would be inappropriate to model a variable with this degree of skew. A variety of transformations could be used to lessen this skew (Box & Cox, 1964). In the previous chapter we used a transformation of the form $ln(x + k)$ to transform the variable x. The k is a small amount added to each value sometimes called the starting value. The square root transformation can also be used to lessen a skewed variable. Here we tried the square root (the R function **sqrt**) of **Final** + **.5**. The histogram in the right panel of Figure 3.2 shows that this transformation reduced the skew.

```
newfinal <- sqrt(Final + .5)
hist(newfinal, main=expression(sqrt(Final + .5)))
par(mfrow=c(1,1))
```

We have used **expression(sqrt(Final + .5))** in the **hist** function to tell R to use mathematical symbols (type **demo(plotmath)** and see Murrell (2006: 97), for a list of symbols that can be printed). The statistics below confirm that the skewness has been reduced.

```
skewness(newfinal)

[1] 0.03587077

skewboot <- boot(newfinal, function(x,i)
    skewness(x[i]),R=1000)
boot.ci(skewboot,type="bca")

BOOTSTRAP CONFIDENCE INTERVAL CALCULATIONS
Based on 1000 bootstrap replicates

CALL :
boot.ci(boot.out = skewboot, type = "bca")

Intervals :
Level               BCa
95%        (-0.3511, 0.4177)
Calculations and Intervals on Original Scale
```

Next we look at the age variable. When making the previous histograms we let R choose how wide to have each of the bars (or bins, as they are sometimes called). There is a lot written on how to choose the number of bars and their widths (Wand, 1997), and within R you can choose one of several algorithms which calculates these. R, however, will not know that the variable **AGEMOS** should be divided into the cultural defined unit years. Therefore, we have to tell R to do this. We tell R to begin new bars at 36, 48, 60, etc. The **seq(36,120,12)** means go from 36 to 120 by 12s (try typing **seq(36,120,12)** in R). The resulting graph is Figure 3.3.

```
hist(AGEMOS, breaks=c(seq(36,120,12)),
    xlab="Age in months")
```

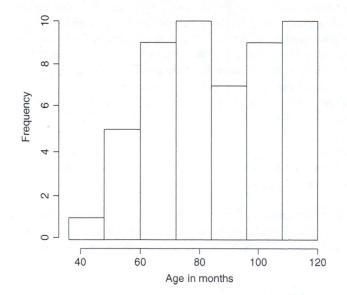

Histogram of AGEMOS

Figure 3.3 A histogram with bin widths of 1 year for the age of children in London et al. (in press)

Now we are ready to do our analyses and to use age to predict **newfinal**. The different analyses will use age defined in four different ways.

1. Age in months (**AGEMOS**) as a numeric variable.
2. Age in years (**AGEYRS**) as a numeric variable.

   ```
   AGEYRS <- trunc(AGEMOS/12)
   ```

   trunc means truncate, so 5.7 years becomes 5 years. This converts this variable into age in years.
3. Age in years (**AGEYRSfac**) as a 6-category factor

   ```
   AGEYRSfac <- as.factor(AGEYRS)
   ```

   The **as.factor** function tells R to read **AGEYRSfac** as a categorical variable (i.e., a factor).
4. Age as a binary variable (**AGE2**)

   ```
   AGE2 <- cut(AGEYRS, br=c(0,6.5,10))
   ```

The **cut** function cuts the variable at the break points, here at the points 0, 6.5, and 10. The 6.5 is because that splits the groups into 4–6 year-olds and the 7–9 year-olds. This will create a binary variable for age.

If we use the first two methods these are simple linear regression between **newfinal** and the age variables. The resulting regressions are:

```
reg1 <- lm(newfinal~AGEMOS)
summary(reg1)
```

```
Call:
lm(formula = newfinal ~ AGEMOS)

Residuals:
   Min          1Q         Median       3Q          Max
  -1.10772    -0.36184    -0.03093    0.36593     1.24272

Coefficients:
             Estimate Std. Error t value Pr(>|t|)
(Intercept) 0.18100    0.32698    0.554    0.582
AGEMOS      0.01776    0.00368    4.825    1.40e-05    ***
---
Signif. codes: 0 '***' 0.001 '**' 0.01 '*' 0.05 '.'
    0.1 ' ' 1


Residual standard error: 0.5448 on 49 degrees of freedom
Multiple R-Squared: 0.3221, Adjusted R-squared: 0.3083
F-statistic: 23.28 on 1 and 49 DF, p-value: 1.404e-05
```

and

reg2 <- lm(newfinal~AGEYRS)
summary(reg2)

```
Call:
lm(formula = newfinal ~ AGEYRS)

Residuals:
   Min          1Q         Median       3Q          Max
  -1.06293    -0.41094    -0.07864    0.40277     1.14544

Coefficients:
             Estimate Std. Error t value Pr(>|t|)
(Intercept) 0.2653     0.3284     0.808    0.423
AGEYRS      0.2150     0.0473     4.545    3.61e-05    ***
---
Signif. codes: 0 '***' 0.001 '**' 0.01 '*' 0.05 '.'
    0.1 ' ' 1


Residual standard error: 0.555 on 49 degrees of freedom
Multiple R-Squared: 0.2965, Adjusted R-squared: 0.2822
F-statistic: 20.66 on 1 and 49 DF, p-value: 3.611e-05
```

The outputs from these regressions are very similar to each other except that the coefficient estimate for the **AGEYRS** regression is about 12 times the size of the one found with the **AGEMOS** regression. This is expected because of the difference in scales between years and months. These two models can be plotted on top of a scatterplot of

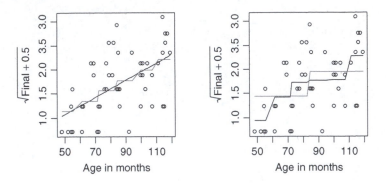

Figure 3.4   Scatterplots comparing the transformed variable for the amount recalled by age in months. The left panel is the model with age treated as a numeric variable in months (*black line*) and years (*gray line*). The right panel is of the models with age treated as a factor (*black line*) and as binary variable (*gray line*)

the data by using the **predict** and the **lines** functions.[3] The **predict** function calculates the predicted values for each of the models, and the **lines** function allows these to be plotted on the scatterplot. In order for this to work the variables have to be sorted according to the *x* axis variable. This variable is sorted with the command **sort(AGEMOS)**. The predicted values have to be in the same order. This is done by telling R to place the predicted values (**predict(reg1)**) in the order of the age variable (**[order(AGEMOS)]**). This has been done in the left panel of Figure 3.4. The **reg1** prediction lines are in black and the reg2 ones are in gray. It is important to note that the scatterplot is in the transformed units. This is printed on the y axis with **ylab=expression(sqrt(Final + .5))**. The black line is when treating age in months and the gray line treating it in years. The lines are similar to each other, although the gray line has a step pattern because all children in the same year band have the same predicted value. The steps are all the same size. This is an assumption of this model.

```
par(mfrow=c(1,2))
plot(AGEMOS,newfinal,ylab=expression(sqrt(Final + .5)),
    xlab="Age in months")
lines(sort(AGEMOS),predict(reg1)[order(AGEMOS)],
    col="black")
lines(sort(AGEMOS),predict(reg2)[order(AGEMOS)],
    col="gray")
```

These analyses were repeated with age as a 6-category factor and a 2-category factor. In these cases the **lm** function runs an ANOVA on the data. We first describe the numeric output and then make graphs depicting the models (right panel of Figure 3.4).

```
reg3 <- lm(newfinal ~ AGEYRSfac)
summary(reg3)
```

---

[3]The function **abline** is only for drawing a straight line which is the entire length of the plot.

```
Call:
lm(formula = newfinal ~ AGEYRSfac)

Residuals:
     Min         1Q          Median      3Q          Max
    -1.05537    -0.23195    0.07367     0.36799     1.15300

Coefficients:
              Estimate  Std. Error  t value  Pr(>|t|)
(Intercept)   0.9391    0.2285      4.109    0.000166    ***
AGEYRSfac5    0.4863    0.3114      1.561    0.125437
AGEYRSfac6    0.7890    0.2891      2.730    0.009017    **
AGEYRSfac7    0.8234    0.2950      2.791    0.007681    **
AGEYRSfac8    0.8395    0.2950      2.846    0.006651    **
AGEYRSfac9    1.3325    0.2891      4.610    3.33e-05    ***
---
Signif. codes:  0 '***' 0.001 '**' 0.01 '*' 0.05 '.'
    0.1 ' ' 1

Residual standard error: 0.5598 on 45 degrees of freedom
Multiple R-Squared: 0.3428, Adjusted R-squared: 0.2698
F-statistic: 4.695 on 5 and 45 DF, p-value: 0.001556
```

The reference category is the 4-year-olds so their mean is 0.9391 utterances. The 5-year olds had $0.9391 + 0.4863 = 1.4254$ utterances, etc. The coefficient estimates allow the means for each year to be calculated. To show these are correct:

`tapply(newfinal,AGEYRSfac,mean)`

```
4            5            6            7            8            9
0.9390518    1.4253104    1.7280871    1.7624776    1.7785640    2.2715376
```

These are the means of the transformed variable.

While the model with **AGEYRS** as a numeric variable assumed the same increase in utterances for each year, this model allows different amounts of increase. Therefore it is able to fit the data better (the R^2 is higher), but the increase is only slight (.2965 to .3428) and it is at a cost. The model is in some sense 5 times more complicated. One way to measure complexity in regressions is the number of coefficients estimated. In the previous models there was only one (the same increase for each year), but there are 5 for the current model. We can compare the fit of these two models with:

`anova(reg2,reg3)`

```
Analysis of Variance Table
Model 1: newfinal ~ AGEYRS
Model 2: newfinal ~ AGEYRSfac
   Res.Df      RSS     Df    Sum of Sq    F         Pr(>F)
1    49     15.0935
2    45     14.1003   4      0.9932      0.7925    0.5363
```

and we see that the increased complexity of the model does not produce a fit that is statistically significantly better, $F(4, 45) = 0.79$, $p = .54$, $\eta_p^2 = .07$. The value for η_p^2 (partial eta-squared) can be calculated from the above output: the difference between the residual sums of squares (RSS) for the two models ($15.0935 - 14.1003 = .9932$) divided by the RSS of the first model, 15.0935.

The final model uses the dichotomized variable **AGE2**. Sometimes it may be useful to dichotomize a variable to present it graphically to friends in a bar, but seldom should it be used outside of drinking establishments (MacCallum et al., 2002). We present it here to illustrate what is implied by the dichotomized method. This could be run as a *t* test, assuming equal variances, as below.

```
t.test(newfinal~AGE2,var.equal=T)

            Two Sample t-test

data: newfinal by AGE2
t = -3.0407, df = 49, p-value = 0.003783
alternative hypothesis: true difference in means is
    not equal to 0
95 percent confidence interval:
    -0.8625846 -0.1761213
sample estimates:
mean in group (0,6.5] mean in group (6.5,10]
              1.430102                1.949455
```

Alternatively, it can be run as a regression. The model being evaluated is the same, but the output is in a different format.

```
reg4 <- lm(newfinal~AGE2)
summary(reg4)

Call:
lm(formula = newfinal ~ AGE2)

Residuals:
    Min         1Q      Median         3Q         Max
  -1.24235   -0.54566   -0.07863    0.41824     1.13275

Coefficients:
                Estimate  Std. Error  t value  Pr(>|t|)
(Intercept)      1.4301      0.1266    11.300   2.98e-15   ***
AGE2(6.5,10]     0.5194      0.1708     3.041    0.00378   **
---
Signif. codes:  0 '***' 0.001 '**' 0.01 '*' 0.05 '.'
     0.1 ' ' 1

Residual standard error: 0.6069 on 49 degrees of freedom
Multiple R-Squared: 0.1587, Adjusted R-squared: 0.1416
F-statistic: 9.246 on 1 and 49 DF, p-value: 0.003783
```

The R^2 now has dropped to 0.16. This means the model predicts the number of utterances less well than the other models. The right panel of Figure 3.4 shows

reg3 and **reg4**. The black line is when the regression treats age as a 6-category factor. The predicted number of utterances goes up in steps, but the steps can be of different heights (unlike **reg2** where we forced the steps all to be the same height). Allowing this flexibility means the model has to estimate many more coefficients than **reg2**. The increase in fit (the R^2 going up slightly) does not justify this increase in complexity. Chapter 7 covers methods to increase flexibility that do not increase complexity as much. The model shown with the gray line is when age has been split into two categories: young and old. The shape of this line shows a problem with the model. The predicted value of utterances is the same for all the children aged between 4 and 6 years old and is the same for all the children above 6 years of age. It assumes that there is some large leap in ability between the ages of 6 and 7. Perhaps there is, but it would be necessary to argue both for the flat parts of the line within the groups and the sudden shift in order to justify this model. If you had such a complex model for the development of children's memories, then the procedures in Chapter 7 would be appropriate for testing it.

```
plot(AGEMOS,newfinal,ylab=expression(sqrt(Final + .5)),
    xlab="Age in months")
lines(sort(AGEMOS),predict(reg3)[order(AGEMOS)],
    col="black")
lines(sort(AGEMOS),predict(reg4)[order(AGEMOS)],
    col="gray")
par(mfrow=c(1,1))
```

LOOKING AT THE DIFFERENCES

The following repeats the analyses conducted above, but for the difference in the number of utterances between the initial and the final interview. This is one method to look at change over time. Alternatives to this are considered in Chapters 4 and 7. The alternatives described later are generally preferred.

A variable is created for the differences between **Final** and **Initial**.

```
Diff <- Final - Initial
```

It is negatively skewed (-0.81) and while in most situations you would transform it (**rank(Diff)** is one possibility), here we will keep with the untransformed variable because we want to compare the results to those in Chapter 4. The four regressions are now re-run. To differentiate them from the previous ones they will be labeled **dreg1** to **dreg4** (for **Diff** regression).

```
dreg1 <- lm(Diff~AGEMOS)
dreg2 <- lm(Diff~AGEYRS)
dreg3 <- lm(Diff~AGEYRSfac)
dreg4 <- lm(Diff~AGE2)
```

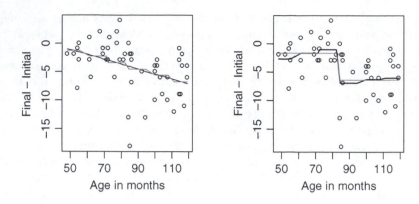

Figure 3.5   Scatterplots between the difference between the number of utterances at the final interview and at the initial interview, with the child's age in months. The left panel shows the models for regression treating age in months (*black line*) and age in years (*gray line*) as linear predictors. The right panel shows the models treating age as a 6-category factor (*black line*) and with age dichotomized (*gray line*). Overall, these show that as age increases the decrease in the amount recalled becomes larger

The **summary** function can be used to look at these. The coefficients all show that there is a negative relationship between age and the difference variable. The following creates Figure 3.5, which is the same as Figure 3.4 for these regressions.

```
par(mfrow=c(1,2))
plot(AGEMOS,Diff, ylab="Final - Initial",
    xlab="Age in months")
lines(sort(AGEMOS),predict(dreg1)[order(AGEMOS)],
    col="black")
lines(sort(AGEMOS),predict(dreg2)[order(AGEMOS)],
    col="gray")
plot(AGEMOS,Diff,ylab="Final - Initial",
    xlab="Age in months")
lines(sort(AGEMOS),predict(dreg3)[order(AGEMOS)],
    col="black")
lines(sort(AGEMOS),predict(dreg4)[order(AGEMOS)],
    col="gray")
par(mfrow=c(1,1))
```

The line for treating age as a 6-category factor is odd. This is because the mean difference goes up slightly between 4 and 5, and between 5 and 6, and then drops a lot between 6 and 7. The means for the difference between final and initial recall for each year are:

```
tapply(Diff,AGEYRSfac,mean)
```

```
     4          5          6          7          8          9
-2.666667  -1.714286  -1.100000  -7.000000  -6.333333  -6.100000
```

These results *suggest* that forgetting is greatest for older kids because the difference in the amount recalled is greatest for them.[4] Using the model with **AGEMOS** as the example, the regression is:

```
summary(dreg1)

Call:
lm(formula = Diff ~ AGEMOS)

Residuals:
    Min          1Q        Median      3Q          Max
  -13.8051     -2.2868     0.6827     2.4205      7.6827

Coefficients:
               Estimate   Std. Error  t value  Pr(>|t|)
(Intercept)     3.06123    2.43603      1.257   0.21484
AGEMOS         -0.08537    0.02742     -3.113   0.00309    **
---
Signif. codes:  0 '***' 0.001 '**' 0.01 '*' 0.05 '.'
    0.1 ' ' 1

Residual standard error: 4.059 on 49 degrees of freedom
Multiple R-Squared: 0.1651, Adjusted R-squared: 0.1481
F-statistic: 9.693 on 1 and 49 DF, p-value: 0.003085
```

So as the number of months goes up, the difference becomes more negative. The equation is:

$$Diff_i = 3.06 - 0.085\ Agemos_i + e_i$$

The standard error is .027 utterances/month, which means an approximate 95% confidence interval is $-.085 \pm 2(.027)$ or $-.30$ to $-.14$. Thus, this shows that as age increases the detriment increases. This *kind of* suggests that older children are more affected by the passage of time, but as we argue in Chapter 4, this conclusion is suspect.

SUMMARY

The purpose of this chapter was to allow you to get used to manipulating a variable and using these in regressions as predictors. The first example showed a oneway ANOVA both with the **aov** function and the **lm** function. The two functions can test the same model, although the format of the output is different. The two procedures, ANOVA and regression, developed relatively separately and are often taught as if they are distinct procedures. It is important to realize that seeing whether group means differ (the purpose often described for ANOVA) is the same as seeing whether there is an association between group membership and the means (the purpose often described for regression).

---

[4]The *suggest* is in italics because this model is based on a dubious assumption, discussed in Chapter 4.

In the second example the predictor variable was one that could be treated in different ways. In R it is straightforward to manipulate a variable and move between numeric and categorical (factor) variable types. The statistical purpose of this example was to stress that information is usually lost when splitting variables into categories. Here, not much is lost changing a variable from being measured in months to being measured in years because for this example there are not huge differences between children's cognitive abilities between months. But, the variable changed greatly when recoded into a young versus old dichotomous variable.

SOME WORDS/CONCEPTS WORTH REMEMBERING

R concepts

* altering variables; truncating and categorizing variables;
* plotting predicted lines; plotting the predicted values of a regression.

R functions

* `as.factor`: reads data as a categorical variable;
* `as.ordered`: reads data as an ordinal variable;
* `tapply`: used to print statistics for different groups;
* `aov`: the function for an ANOVA;
* `plot`: makes lots of different kinds of figures;
* `anova`: compares two or more regression objects;
* `contrasts`: how to find or change contrasts;
* `dim`: how to find or change the dimensions of data;
* `expression`: used for writing mathematical expressions;
* `cut`: used to cut variables in categories;
* `sort`: sorts a variable from low to high;
* `order`: finds the order of a variable (useful with `lines`);
* `predict`: the predicted values from a regression.

Statistical concepts

* ANOVA: is a type of regression;
* dichotomization: is a poor scientific strategy and should be avoided.

FURTHER READING

The classic reference within psychology on ANOVA being a regression with categorical variables is:

Cohen, J. (1968). Multiple regression as a general data-analytic system. *Psychological Bulletin*, *70*, 426–443. Here is a good quotation from it:

If you should say to a mathematical statistician that you have discovered that linear multiple regression analysis and the analysis of variance (and covariance) are identical systems, he would mutter something like, 'Of course—general linear model,' and you might have trouble maintaining his attention. If you should say this to a typical psychologist, you would be met with incredulity, or worse. (p. 426)

The classic reference for not dichotomizing variables is:

Cohen, J. (1983). The cost of dichotomization. *Applied Psychological Measurement, 7,* 249–253.

A more recent and more thorough account is in:

MacCallum, R.C., Zhang, S., Preacher, K.J., & Rucker, D.D. (2002). On the practice of dichotomization of quantitative variables. *Psychological Methods, 7,* 19–40.

4

ANCOVA: Lord's paradox and mediation analysis

Learning outcomes

1. Conducting and interpreting ANCOVA.
2. Multiple regression.
3. Producing a complex graph.
4. Writing functions in R.

Analysis of Covariance (ANCOVA) is a popular technique. It is often taught as an extension of ANOVA, but it can also be taught as multiple regression. Here we treat it as a multiple regression. This allows us to introduce multiple regression, which is used in several subsequent chapters. The idea behind ANCOVA is that one variable, called the covariate, is used to predict the response variable, and then you see how well a second predicter is able to improve the model beyond using the covariate alone. In practice you can have several covariates and look for any additional predictive value in a large number of other variables, but we will keep with the situation where there are only three variables: one response variable, one covariate, and one additional predictor variable.

When ANCOVA is taught as an extension of ANOVA, it is often assumed that the covariate is a continuous numeric variable and that the second predictor is a factor. When treated as a type of regression these assumptions are not necessary. If the variable is type **numeric** then the **lm** function in R treats it appropriately, and if it is type **factor** then the function also treats it appropriately. The ANCOVA model compares two regressions:

$$\text{model 1} \quad y_i = \beta 0 + \beta 1 \ covariate_i + e_i$$
$$\text{and} \quad \text{model 2} \quad y_i = \beta 0 + \beta 1 \ covariate_i + \beta 2 \ x_i + e_i$$

The difference between the fit of these two models shows how important x_i is in predicting y_i after the covariate's influence has been taken into account. This is often described as the effect of x_i on y_i after partialling out the covariate.[1]

Two situations are used to present ANCOVA. The first is where you are trying to measure change between two points in time and you want to see if two (or more) groups differ. We will use the London et al. (in press) data set described in Chapter 3. The ANCOVA produces a result that looks at odds with what we found in Chapter 3. This is an example of Lord's Paradox (1967). The second situation is where you have experimentally manipulated one variable and you want to see if its effect on another variable is due to its effect on a third variable. This is called *mediation analysis*.

LORD'S PARADOX

Lord (1967) described a fictitious example where two statisticians, faced with the same data set and the same basic research question, came to different answers. The research question was whether there are group differences on some measure at time 2 after taking into account values at time 1. The approaches the statisticians took were an ANOVA on the differences between the scores (time 2 – time 1) and an ANCOVA on time 2 scores partialling out time 1 scores. These are the two most popular approaches that psychologists continue to use in this situation. Because both of these approaches appear to address the same substantive question and yet can produce different conclusions, this phenomenon has become known as Lord's Paradox.

Much has been written on Lord's Paradox in the statistics literature (e.g., Wainer, 1991). Hand (1994) described how Lord's Paradox is only paradoxical because people are not precise enough about their hypotheses. Thus, if London et al.'s (in press) hypothesis was whether the ages differed in the *difference* in the amount recalled at the points in time, then this would translate into subtracting the amount recalled at the initial interview from the amount recalled at the final interview, and seeing whether this difference is associated with the child's age. This, the ANOVA method, was advocated by one of Lord's (1967) statisticians, and is what was done in Chapter 3. Lord's other statistician suggested an ANCOVA with the scores from the initial interview partialled out. Formal comparison of the ANOVA on the change scores and the ANCOVA have led many statisticians to argue that if you are interested in whether the grouping variable is causing a difference in the response variable at time 2, then the ANCOVA approach is usually, but not always, preferred (Wainer, 1991; Wright, 2006b).

The two approaches can be written as follows:

- ANOVA: $Final_i = Initial_i + \beta 1\ Age_i + \beta 0 + e_i$
- ANCOVA: $Final_i = \beta 2\ Initial_i + \beta 1\ Age_i + \beta 0 + e_i$

---

[1] Several other phrases are used for this. 'Covarying out the covariate' has the same meaning, but most dictionaries will not have the word 'covarying' in them. 'After taking the covariate into account' is okay, but it does not tell you how the covariate has been taken into account. 'After controlling for the covariate' suggests that the researcher has done something to the covariate, and a common reason for doing an ANCOVA is that the researcher cannot manipulate the covariate. 'After partialling out the covariate' is not a perfect phrase, but it seems the best of these.

Viewed in this way it is clear that in one sense (at an algorithmic level) the only difference between the approaches is that with the ANOVA approach $\beta2$ is assumed to be 1, and with the ANCOVA approach it is estimated. While these are often referred to as the ANOVA and ANCOVA approaches, as shown in Chapter 3 and repeated here, both of these approaches can be modeled with the **lm** function.

Example 5 – Children's forgetting 2

- Data: From London et al. (in press, Exp. 2).
- Research question: Taking into account the initial amount recalled, how is age related to final recall?
- Purpose: To get accustomed to using multiple regression and to learn about Lord's Paradox.

The data are accessed in the same way as in Chapter 3.

```
webreg <- "http://www.sagepub.co.uk//wrightandlondon//"
lordex <- read.table(paste(webreg,"lordex.txt",sep=""), header=T)
attach(lordex)
```

In Chapter 3 the difference between the number of utterances at the final interview was subtracted from the initial interview and a regression was run on this difference. The ANCOVA approach involves predicting the **Final** from **Initial** and seeing whether the age variable is able to help predict **Final** after partialling out the **Initial** scores. We will use the variable **Final** rather than the transformed variable. This allows comparisons with the methods used in Chapter 3 and those used later in Chapter 7. Using either **AGEMOS** (age in months) or **AGEYRS** (age in years) produce nearly identical results. We will use the **AGEMOS** variable for the regressions, but use both of them in constructing a graph. Notice that all of these variables are numeric.

ANCOVAs should be conducted in three steps. First you should see how well the covariate (or covariates) predicts the response variable. Most of the time the covariate is associated with the response variable, but this is not always the case. Next you should see whether adding the other predictor variable increases the fit of the model. Finally, you should look at the interaction between this variable and the covariate. This tests whether any effect of predictor variable depends on the value of the covariate. You may have several covariates and several other predictor variables, and interactions among all of these. If you have several covariates and predictor variables each of these steps will have several parts and you should proceed carefully entering variables (see Chapter 5 for related issues). These three steps are done with the following commands (**Initial*AGEMOS** includes the interaction of **Initial** and **AGEMOS** and both of these variables' main effects). The **summary** function is used with each to extract important statistics.

```
lm1 <- lm(Final ~ Initial)
summary(lm1)

Call:
lm(formula = Final ~ Initial)
```

```
Residuals:
   Min        1Q    Median       3Q       Max
-2.6985   -1.7312   -0.2614   1.1294   5.5201

Coefficients:
            Estimate Std. Error t value Pr(>|t|)
(Intercept) 1.29404    0.50690    2.553   0.013850 *
Initial     0.21859    0.05818    3.757   0.000457 ***
---
Signif. codes: 0 '***' 0.001 '**' 0.01 '*' 0.05 '.' 0.1 ' ' 1

Residual standard error: 2.053 on 49 degrees of freedom
Multiple R-Squared: 0.2237,  Adjusted R-squared: 0.2078
F-statistic: 14.12 on 1 and 49 DF, p-value: 0.0004571
```

lm2 <- lm(Final ~ Initial + AGEMOS)
summary(lm2)

```
Call:
lm(formula = Final ~ Initial + AGEMOS)

Residuals:
   Min        1Q    Median       3Q       Max
-3.4304   -0.8374   -0.4666   0.7103   5.1568

Coefficients:
            Estimate Std. Error t value Pr(>|t|)
(Intercept) -1.72927    1.21740   -1.420   0.16193
Initial      0.10525    0.06901    1.525   0.13379
AGEMOS       0.04441    0.01645    2.699   0.00956 **
---
Signif. codes: 0 '***' 0.001 '**' 0.01 '*' 0.05 '.' 0.1 ' ' 1

Residual standard error: 1.933 on 48 degrees of freedom
Multiple R-Squared: 0.326,  Adjusted R-squared: 0.2979
F-statistic: 11.61 on 2 and 48 DF, p-value: 7.723e-05
```

lm3 <- lm(Final ~ Initial*AGEMOS)
summary(lm3)

```
Call:
lm(formula = Final ~ Initial * AGEMOS)

Residuals:
   Min        1Q    Median       3Q       Max
-3.4185   -0.7978   -0.4959   0.7168   5.1251
```

```
Coefficients:
                  Estimate Std. Error t value Pr(>|t|)
(Intercept)     -1.9960325  2.2986626  -0.868   0.390
Initial          0.1476360  0.3163338   0.467   0.643
AGEMOS           0.0478716  0.0301855   1.586   0.119
Initial:AGEMOS  -0.0004931  0.0035896  -0.137   0.891

Residual standard error: 1.953 on 47 degrees of freedom
Multiple R-Squared: 0.3263,  Adjusted R-squared: 0.2833
F-statistic: 7.587 on 3 and 47 DF,  p-value: 0.0003078
```

There are several measures that can be used to compare the fit of these models, and more discussion about these in Chapter 5. For now, see that the R^2 increases from 0.2237 to 0.3260 to 0.3263 from **lm1** to **lm2** to **lm3**. So, the increase is substantial when the main effect of **AGEMOS** is added, but is minimal when the interaction is added. Hypothesis tests of these differences are calculated with the **anova** function.

anova(lm1,lm2,lm3)

```
Analysis of Variance Table

Model 1: Final ~ Initial
Model 2: Final ~ Initial + AGEMOS
Model 3: Final ~ Initial * AGEMOS
  Res.Df     RSS Df Sum of Sq      F  Pr(>F)
1     49 206.531
2     48 179.309  1    27.222 7.1382 0.01034 *
3     47 179.237  1     0.072 0.0189 0.89132
---
Signif. codes: 0 '***' 0.001 '**' 0.01 '*' 0.05 '.' 0.1 ' ' 1
```

η_p^2 values can be calculated from the RSS values, so η_p^2 for **AGEMOS** partialling out **Initial** is $(27.222)/206.531 = .13$. The first comparison shows that model 2 is a significantly better fit than model 1 ($F(1, 48) = 7.14$, $p = .01$, $\eta_p^2 = .13$). The second comparison shows that the interaction between **AGEMOS** and **Initial** for predicting **Final** is non-significant ($F(1, 47) = 0.02$, $p = .89$, $\eta_p^2 = .00$). Thus, **lm2** looks the best of these models. From the output above, the regression equation is:

$$Final_i = -1.73 + 0.105\ Initial_i + 0.044\ AGEMOS_i + e_i$$

This is the basic ANCOVA. This is exciting (for us at least). The coefficient for age is positive. It shows that as age increases so does the predicted value for recall at the final interview. This is in the opposite direction than that suggested by the analysis in Chapter 3. This is an example of Lord's Paradox.

Some graphs are necessary to understand any data set. Here the **lattice** library (Sarkar, 2008) is used to make trellis graphs. Many statisticians think trellis graphs are incredibly useful. A very simple one is done in Figure 4.1. It draws the scatterplot for each age-year group.

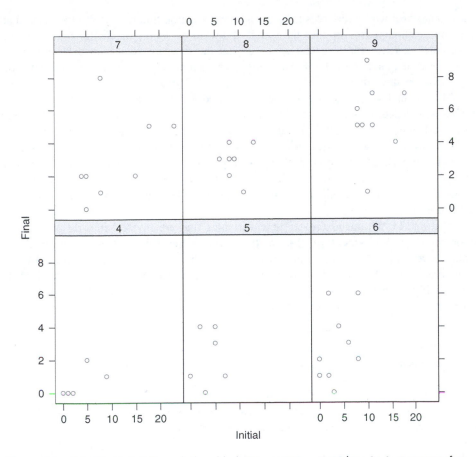

Figure 4.1 A trellis plot of the relationship between **Final** and **Initial** scores for the six age groups

```
library(lattice)
AGEYRS <- trunc(AGEMOS/12)
scatmat <- xyplot(Final~Initial | as.factor(AGEYRS))
print(scatmat)
```

From this graph is it clear that there is less variability for some ages than others. Here are standard deviations.

```
tapply(Initial, AGEYRS, sd)

     4         5         6         7         8         9
3.311596 2.299068 2.998148 6.964194 2.147350 3.314949


tapply(Final, AGEYRS, sd)

     4         5         6         7         8         9
0.836660 1.772811 2.043961 2.500000 1.201850 2.538591
```

For many real world variables (like income and reaction times, etc.) it is expected that as the mean increases so does the standard deviation. As the means for both of these variables increase with age you would expect the standard deviations to do this also. This does not happen for the 8 year-olds, so we might expect any regressions for this group to be unreliable.

More complex figures can also be drawn, and an example is shown in Figure 4.2. This is included to show some of the possibilities in R. This shows both the ANOVA on the differences from Chapter 3 and the ANCOVA from this chapter. Complex graphs like this can take a long time to make, but it does get quicker once you get used to whichever package that you use. You probably do not want to spend days on this graph, so just copy the code below to show what you get (the code is on the book's web page). This is a fairly detailed example so is included for you to look through and as an illustration.

The **split** command creates new variables that are divided into different parts for each age. So, **sexit[[1]]** are the initial scores for the youngest year group and **sexit[[2]]** are the values for the second youngest group. **predmod4** are the predicted values for the model which allows the slopes to vary for each year group. If you want parallel slopes replace the **Initial*as.factor(AGEYRS)** with **Initial+as.factor(AGEYRS)**.

```
sexit <- split(Initial,AGEYRS)
sfoll <- split(Final,AGEYRS)
predmod4 <- split(lm(Final~Initial*as.factor(AGEYRS))
   $fitted.values,AGEYRS)
plot(Initial,Final,pch=19,cex=.5,col="black",
   xlab="Initial interview", ylab="Final interview", xlim=c(0,25),
   ylim=c(0,10), cex.lab=1.3, las=1, font.lab=1.5)
```

When the function **abline** has two points entered it assumes they are the coefficients of a line $y = ax + b$. Here the line drawn is $y = 1x + 0$, which is a diagonal line through the origin. This shows where the values of **Initial** equal those of **Final**. Next, we want to draw a line for each of the six age years. The code **for (i in 1:6)** tells the computer that you want it to run the remainder of the line (or whatever is included within **{ }** if it is a longer set of commands) six times and put the numbers 1 to 6 in where **i** is for each one. So it runs **lines(sexit[[1]],predmod4[[1]], col = "black")** and then **lines(sexit[[2]],predmod4[[2]], col = "black")**, and so on up to 6. The **for** function is often used in R to save having to write similar sets of commands over and over. We have used the **lines** command rather than **abline** because we wanted the lines to only go as far as the observed data in each group (i.e., not to extrapolate beyond the data). It is worth noting that we have not worried about using the **sort** or **order** functions here as we did in Chapter 3. This is because the lines are straight. For regressions covered in later chapters sorting would be necessary.

```
abline(0,1,lty = 3)
for (i in 1:6) lines(sexit[[i]],predmod4[[i]],
   col="black")
```

The following calculates the year means and then connects them with a thick (**lwd=4**, **lwd** stands for line width and the default is 1) **"gray"** line. The **lend="round"** tells R to make rounded corners where the lines change directions. This option is only necessary when thick lines are used.

```
meanexit <- c(mean(sexit[[1]]),mean(sexit[[2]]),
    mean(sexit[[3]]), mean(sexit[[4]]),
    mean(sexit[[5]]),  mean(sexit[[6]]))
meanfoll<- c(mean(sfoll[[1]]),mean(sfoll[[2]]),
    mean(sfoll[[3]]),
    mean(sfoll[[4]]), mean(sfoll[[5]]), mean(sfoll[[6]]))
lines(meanexit,meanfoll,col="gray",lwd=4,lend="round")
```

Adding text to a graph is important so it is worth repeating how this done. In R you put in the x and y coordinates, then in **" "** you put the text. The argument **pos** defines whether it is left, right, top, or bottom justified. Some trial and error is usually required to get the text where you want it. If you want to include values of variables or odd characters use the **paste** or **expression** functions. These are illustrated in many graphs throughout this book. The numbers in the **arrows** function are for the start and the end of the arrow, **length** is for the arrow head. You can put in multiple elements in the **text** command, as below, just make sure there are the same number of x and y values and objects to plot. The **\n** in the **text** functions tells R to put in a line break.

```
text(12,10,"Final = Initial",pos=4)
arrows(12,10,10,10,length=.1)
text(c(13,21,12.1,0,0,4.4),c(5.5,4.5,2.2,3.7,1.5,.5),
    c("9","7","8","6","5","4"))
points(meanexit,meanfoll,pch=19)
lines(c(14,16.7), c(9.2,9.2),col="gray",lwd=4)
text(16.7,9.2,"Age group means", pos=4)
text(14,8.5,
    "Means increase with age\nfor both interviews",pos=4)
text(-.3,9.5,
    "Intercepts higher for\nolder children\n (Ancova)",pos=4)
text(14,1,
    "Older children further away\nfrom dashed line (Anova).",
    pos=4)
```

What this shows is that for all the groups except for 8-year olds there is a positive relationship between scores on the two interviews (i.e., all the thin black lines except the one for 8-year olds have positive slopes). The basic ANCOVA approach, which assumes the lines are parallel, seems adequate. Further, the intercepts for all these groups tend to be higher for the older children. This is what the ANCOVA is testing: for any initial interview score the predicted score at the final interview is higher for older children.

For the ANOVA approach in Chapter 3, the estimate for the age coefficient was negative, meaning as age increases the detriment from increased elapsed time also increases. In other words, the average decrease in the amount reported is larger for the older children than for the younger children. This is shown by the means for the older age groups being further away from the dashed line, $Final_i = Initial_i$, made with **abline(0,1,lty = 3)** in Figure 4.2. With the ANCOVA approach, conducted

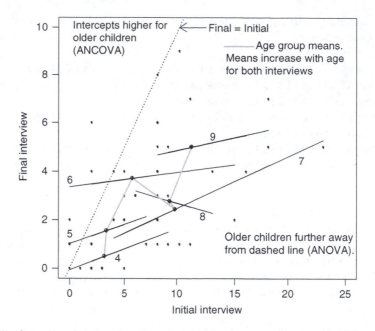

Figure 4.2   A scatterplot showing the relationships between final and initial recall for the 6 age groups from London et al (in press)

in this chapter, the estimates for age are positive. This means that, controlling for initial scores, the older children have higher predicted final scores. As these appear to be the opposite conclusions, it logically must mean that they address different questions.

Let's imagine two scenarios of forensic importance which relate to these data. Suppose a crime has occurred and there are two witnesses: a 5- and an 8-year old. For the first scenario, suppose that you have limited resources and that you can interview one of the children soon after the crime, but you have to wait ten months to interview the other child. You want to get the most information possible from the combination of the two children. Which child should you interview first? Because the ANOVA says that the decrease is greatest for older children you would interview the 8-year old first, and wait for the 5-year old. It is as if the younger child's recall is less affected by the passage of time.

For the second scenario, suppose that the two children were both interviewed soon after the crime and recalled the same amount at the initial interview. One year later you want to call one of the children to the witness stand during the trial. The ANCOVA approach showed that, controlling for initial recall, older children do better. Thus, you would interview the older child. It is as if the older child's recall is less affected by the passage of time.

As descriptions of the data, both of these approaches are valid, but they apply to different scenarios. Each of Lord's original statisticians were correct, but for different situations. These situations are both fictional, and while we can imagine situations like them, most scientists will be interested in the causal inference suggested by each. Because the conclusions are contradictory, both cannot be valid. These data provide a striking example of Lord's Paradox. Sometimes the choice of a statistical procedure may determine which side of the magical $p = .05$ a solution may lie on. But here the two methods give large statistically significant effects in *opposite* directions! Taken at face value, the ANOVA approach suggests that delay exerts a more detriment effect on older children's reports, while the ANCOVA approach suggests the opposite. Which analytic method accounts more accurately for these data?

When Lord (1967) introduced this paradox he did not say which approach was generally preferred. It was not until Rubin's (1974) model of causal inference was applied to Lord's Paradox (Holland & Rubin, 1983) that it was shown when each of these approaches is more likely to be valid for causal inference. Rubin's model is usually described in relation to experimental groups and causation, but it can also be used when the interest is in a quasi-experimental group, like age (Wainer, 1991). When the variable is quasi-experimental, interest is usually in an association rather than a cause. Wainer and others have argued that for the ANOVA approach to be appropriate you have to be able to assume that, as time progresses, people will recall the same amount of information. The ANCOVA approach does not make this assumption, so is preferred for inference here since we expect people to remember less as time elapses (as evident from 125 years of memory research). The ANCOVA assumes the amount recalled goes down linearly with time. From memory research we know that the memory decay function is more complex, and this complexity could be built into the analysis, but for most purposes the linear decay assumption of the standard ANCOVA is adequate. A more flexible way of exploring these data is discussed in Chapter 7.

Answer: Usually the ANCOVA approach is preferred to running an ANOVA on the change or difference scores. See Wainer (1991) and Wright (2006a) for more details.

An alternative way to argue against the ANOVA method, here, is by floor effects and 'levels of measurement.' The floor effects argument is simple. As can be seen in Figure 4.2, the younger children cannot recall that much less at the second testing than at the earlier testing because they did not recall that much at the first testing. If you only recall 1 thing at the first testing, you can only recall 1 less at the 10 month testing. This is a possible floor effect. There is controversy surrounding how and if to use 'levels of measurement' when choosing statistical procedures (Lord, 1953; Velleman & Wilkinson, 1993; Wright & London, 2009). If levels of measurement is taken at face value, and ANOVA is used, the implicit assumption is that the difference between recalling 2 items at the initial testing and 1 at the 10 month testing is the same amount (in some psychological sense) as the difference between recalling 12 and 11 items. This does not seem valid. Another option would be to use the ratio of these numbers (or other transformations), so equating a drop from 2 to 1 as the same as from 12 to 6, but this also presents problems. The ANCOVA model implies a more complex relationship, which is more realistic here. It is easy to use 'levels of measurement' to argue *against* doing a particular test, but it is hard to use it to argue *for* any test. This is one reason why we urge researchers not to treat 'levels of measurement' as a restrictive doctrine, but more as a guide (Wright & London, 2009).

THREE MISINTERPRETATIONS OF ANCOVA

There are three mistakes often made in interpreting ANCOVAs. The first is to make your interpretation as if you did not have any covariates, and interpret the result as if it was just an ANOVA. Here it would be to use the ANCOVA to say that as age went up the amount recalled went up. This is not the correct interpretation because you have to include the statement that it is after partialling out the covariate. The second common misinterpretation is that once you have partialled out one covariate (or some covariates) that you can make causal conclusions without further assumptions. This is not true, you still need to be careful. It is easier to make causal conclusions with experimental data (Wright, 2006b). If you do not have experimental data then ANCOVA can help, but no statistical procedure will allow causal conclusions without assumptions.

The final misinterpretation is 'throwing out the baby with the bath water.' If you were interested in children's mathematical attainment and covaried out an IQ measure, you would be removing variance related to what you are interested in. This would have to be reflected in your conclusions. It is worth always thinking about ANCOVA in the three steps described in the text, graphing the results, and carefully stating your conclusions.

MEDIATION ANALYSIS AND WRITING FUNCTIONS

Particularly within social and health/clinical psychology, many researchers are interested in what are called *mediator* variables (MacKinnon et al., 2007). There is a similarity between simple mediation analysis and Lord's Paradox because each involves just three variables and uses ANCOVA to try to draw out some type of causal inference. However, time plays a different role in these. In Lord's Paradox the assumption is one variable is measured at time 1, there is a manipulation, and then the same variable is measured again at time 2. The manipulation cannot affect the initial measurement. In mediation analysis, there is a manipulation, then the researchers measure the potential mediator and some outcome variable. These are often measured simultaneously, but it is useful to conceptualize the mediator as occurring before the outcome variables. This will become clear through an example.

Let's suppose that we are interested in sun block use. We select a random sample of 100 people who are planning on having a weekend on Miami beach. A random 50% are sent a leaflet with some facts about skin cancer and are asked to imagine what it is like to have skin cancer. The other half receives nothing. Call the experimental variable **leaflet** with the value 0 for the control and 1 for the leaflet condition. When the people arrive at the beach they are asked several questions, including what they believe is the likelihood that they will get skin cancer at some point in their lives (**likely** on a 0 to 10 scale, where 10 means very likely), and whether they were planning on using sun block (**plan** on a 1 to 7 rating scale, with 7 meaning definitely). Here are some hypothetical data. Notice that when we create the data we have a variable for **fairskin**. There are many variables that are associated with sun block use, and most of these will not be measured. We will assume that we have not measured this for our analyses.

```
set.seed(143)
leaflet <- rep(c(0,1),each=50)
fairskin <- rbinom(100,1,.5)
likely <- rbinom(100,10,.20 + .2*leaflet + .2*fairskin)
plan <- rbinom(100,7,likely/15+leaflet*.2)
```

We have two research questions. First, does the leaflet we sent out increase the likelihood that people plan to use sun block? Second, if there is an effect, is any part of this effect due to increasing how likely it is that people think that they will get cancer? The first question can be addressed easily with a standard regression. The second question is more complex and requires that we run series of regressions each asking a different question. These can be done directly with the **lm** function. Here they are with some of the relevant output. The **summary** function on its own would have produced more output. Including the **$coef** at the end means just the part relevant to the coefficients is printed.

Does **leaflet** *predict* **likely**? *YES*

summary(lm(likely~leaflet))$coef

	Estimate	Std. Error	t value	Pr(>|t|)
(Intercept)	3.00	0.2593870	11.565729	5.059162e-20
leaflet	1.92	0.3668287	5.234051	9.450470e-07

Does **leaflet** *predict* **plan**? *YES*

summary(lm(plan~leaflet))$coef

	Estimate	Std. Error	t value	Pr(>|t|)
(Intercept)	1.60	0.1878341	8.518157	1.964932e-13
leaflet	2.08	0.2656375	7.830221	5.830890e-12

Does **likely** *predict* **plan**? *YES*

summary(lm(plan~likely))$coef

	Estimate	Std. Error	t value	Pr(>|t|)
(Intercept)	0.2972691	0.25340697	1.173090	2.436032e-01
likely	0.5915987	0.05680571	10.414423	1.528473e-17

Does **leaflet** *predict* **plan**, *after partialling out* **likely**? *YES*

summary(lm(plan~likely+leaflet))$coef

	Estimate	Std. Error	t value	Pr(>|t|)
(Intercept)	0.2375880	0.22615226	1.050566	2.960681e-01
likely	0.4541373	0.05727007	7.929750	3.783014e-12
leaflet	1.2080563	0.23525108	5.135179	1.453311e-06

The first regression (which is the same as a t test) shows that the leaflet increases the likelihood that people think they will get skin cancer ($t(98) = 5.23, p < .001$) with the mean increasing from 3.00 to 4.92. The second (also a t test) shows that the leaflet increases people's plans to use sun block ($t(98) = 7.83, p < .001$) with the mean increasing from 1.60 to 3.68. The third regression shows that the variables **likely** and **plan** are highly associated (we are careful to avoid causal terms here: $t(98) = 10.41$, $p < .001$). The final model shows that after controlling for people's believed likelihood of getting cancer, the leaflet still had an effect ($t(97) = 5.14, p < .001$). This is the standard ANCOVA and means that at least part of the effect of the leaflet cannot be accounted for by changing the belief about the likelihood of getting cancer. The question is: can any of the leaflet effect be accounted for by this belief. In statistical jargon, is **likely** a partial mediator of the **leaflet** effect? The most common test of this is the Sobel test (MacKinnon et al., 2007). If we let a be the effect of **leaflet** on **likely** (1.92, se $= .37$), b be the effect of **likely** on **plan** after taking into account **leaflet** (0.45, se $= .06$), then the effect associated with the path from **leaflet** through **likely** to **plan** is $a*b$ with a standard error of the square root of $(se_a^2*b^2 + se_b^2*a^2 + se_a^2*se_b^2)$.[2] The Sobel test is the ratio of these and is usually assumed to be normally distributed. Here this is:

```
(1.92*.45)/sqrt(.37^2*.45^2 + .06^2*1.92^2+.37^2*.06^2)
4.24
```

It is always a problem calculating statistics in this manner because you can make typing errors, there will be rounding errors (in the equation we square .06 which gives us .0036, but because the real value is closer to .057, a better square would be .0032, a 10% difference), and it is not fun typing lots of numbers. So, you could write a function for this. A simple example is:

```
sobel1 <- function(a,b,sea,seb){
sobel <- (a*b)/sqrt(sea^2*b^2+seb^2*a^2+sea^2*seb^2)
paste("Sobel z =", format(sobel,nsmall=2,digits=3))}
```

The command **format(sobel,nsmall=2,digits=3)** tells R to print at least 3 digits for the value and at least 2 after the decimal point. This prevents R from printing too many digits. Then each time you want to run the Sobel test you could just enter the coefficients and their standard errors. For this example you would write:

```
sobel1(1.92,.45,.37,.06)
```

and you get the same as above:

```
"Sobel z = 4.24"
```

This still includes rounding errors, so increasing the precision by one digit we get:

```
sobel1(1.920,.454,.367,.057)
```

---

[2] The '$se_a^2*se_b^2$' is not included in some sources because it is usually very small. Some researchers suggest subtracting it rather than adding it, but given that it is usually very small, detailed discussion of this is not warranted.

```
[1] "Sobel z = 4.35"
```

While this saves one calculation, we would still have to do all the regressions, and if we still thought *p* values were of use, we would need to look that up. We might even want to draw a graph of the effects. This suggests that it might be worth writing a function that does all this. A function that does this is shown in the box below. You can copy it from there, or run it with the following command which reads it from the book's web page:

source("http://www.sagepub.co.uk//wrightandlondon//mediator.R")

This function does what is often called simple mediation. Type **mediator** with the name of the experimental variable, the outcome variable, and the mediator variable in order. For example:

mediator(leaflet,plan,likely)

produces the following output

```
[1] "Sobel z = 4.34 , p = 1.4e-05"
```

which is Sobel's *z* and the associated *p* value. Notice that this new figure, *z* = 4.34, is very similar to the number found earlier. The difference is that more precise values are used here so there is less rounding error. The function also produces the graph shown in Figure 4.3.

The **mediator** function is shown in the box below. As with the other figures in this book, to reproduce the figure exactly may require changing the size of the graphics window. This function is for simple mediator analyses. Variations can be added to it, for example using bootstrap procedures rather than relying on the asymptotic *p* value, which is known to have a fairly large error associated with it (so should be viewed as the approximate *p* value). Only two new functions are used: **rect** and **invisible**. **rect** draws a rectangle if you provide is with the four values to define the two corners of the rectangle. Here it is used to make multiple rectangles. The **invisible** function

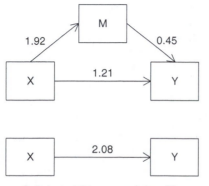

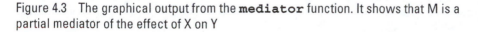

Figure 4.3   The graphical output from the **mediator** function. It shows that M is a partial mediator of the effect of X on Y

is used so the **mediator** function does not print to screen Sobel's z and its associated p value to many decimal points, but that these values are saved if you type **values <- mediator(x,y,z)**.

Here is the **mediator** function:

```
## Simple Mediation Function with Figure of effects
mediator <- function(x,y,m, ...){
  # Put in experimental variable x, outcome variable y, then mediator m.
  # Some of this is written so that it works in a mediator
  # function for more complex problems.
  reg0 <- lm(y~x)
  reg1 <- lm(m~x)
  reg2 <- lm(y~m+x)
  c <- summary(reg0)$coefficients[2,1]
  sc <- summary(reg0)$coefficients[2,2]
  a <- summary(reg1)$coefficients[2,1]
  sa <- summary(reg1)$coefficients[2,2]
  b <- summary(reg2)$coefficients[2,1]
  sb <- summary(reg2)$coefficients[2,2]
  cp <- summary(reg2)$coefficients[3,1]
  scp <- summary(reg2)$coefficients[3,2]
  sobel <- (a*b)/sqrt(b^2*sa^2 + a^2*sb^2 + sa^2*sb^2)
  psobel <- format(2*(1-pnorm(abs(sobel))),digits=2,nsmall=2)
  plot(c(0,100),c(0,110),col="white",ann=F,tck=0,col.axis="white")
  rect(c(10,10,70,70,40),c(10,50,10,50,80),c(30,30,90,90,60),c(30,70,30,70,100))
  arrows(c(30,30,20,60),c(20,60,70,90),c(70,70,40,80),c(20,60,90,70),length=.15)
  text(c(20,20,80,80,50),c(20,60,20,60,90),c("X","X","Y","Y","M"),cex=2)
  text(30,80,paste(format(a,digits=2,nsmall=2)),pos=2,cex=1.3)
  text(70,80,paste(format(b,digits=2,nsmall=2)),pos=4,cex=1.3)
  text(50,20,paste(format(c,digits=2,nsmall=2)),pos=3,cex=1.3)
  text(50,60,paste(format(cp,digits=2,nsmall=2)),pos=3,cex=1.3)
  text(50,2,paste("Sobel z = ",format(sobel,digits=2,nsmall=2),"; p = ",
    psobel),cex=1.3)
  print(paste("Sobel z =", format(sobel,digits=3,nsmall=2),", p =", psobel))
  invisible(c(sobel,psobel))
}
```

SUMMARY

ANCOVA is often taught as an extension to ANOVA. This is a shame because it is really just a multiple regression where there is an emphasis on the order in which the predictor variables are input into the model. If it were taught as a way to introduce multiple regression it would make the similarity between ANCOVA and regression clearer, would allow people to realize that categorical and/or continuous variables could be used with either the covariates or the predictor variables, and it would make people doing multiple regressions more careful about how they describe their results. When marking multiple regression assignments, one of the most annoying things is when it is not clear that the person understands that the effect estimated is conditional on all of the other predictor variables. This will be discussed further in Chapter 5.

There are several different examples of ANCOVA that could have been used to illustrate the technique. We chose showing an example of Lord's Paradox, because that stresses the importance of choosing the best statistical procedure, and mediation analysis, because this is often required when trying to tease apart different effects.

SOME WORDS/CONCEPTS WORTH REMEMBERING

R concepts

- lattice library: an R library for graphs in grids;
- more on graphs: further options for the scatterplot;
- plotting predicted lines: plotting the predicted values of a regression;
- writing functions: a simple function and a complex function are shown.

R functions

- `xy.plot`: scatterplots for groups in lattice library;
- `for (i in 1:k)`: a loop function in R;
- `/n`: line breaks in text function;
- `arrows`: prints arrows on graphs;
- `$coef`: how to access regression coefficients;
- `format`: controls the number of digits printed;
- `source`: accesses functions;
- `rect`: draws rectangles;
- `invisible`: so returned values are not printed to the screen.

Statistical concepts

- ANCOVA: is a type of multiple regression;
- Lord's Paradox: when looking at change, usually use ANCOVA;
- Mediation analysis: to test if an effect is due to a mediating variable.

FURTHER READING

The key reference on Lord's Paradox, and an enjoyable read, is:

Lord, F. M. (1967). A paradox in the interpretation of group comparisons. *Psychological Bulletin*, 72, 304–305.

A more recent and detailed paper is:

Wainer, H. & Brown, L. M. (2004). Two statistical paradoxes in the interpretation of group differences: Illustrated with Medical School Admission and Licensing Data. *American Statistician*, *58*, 117–123.

An excellent review of mediation:

MacKinnon, D. P., Fairchild, A. J. & Fritz, M. S. (2007). Mediation analysis. *Annual Review of Psychology*, *58*, 593–614.

A good introduction to the area:

Rutherford, A. (2000). *Introducing ANOVA and ANCOVA: A GLM approach.* London: Sage.

5

Model selection and shrinkage

Learning outcomes

1. Different methods for selecting a simpler model from a large set of predictors;

   - best subset regression
   - ridge regression
   - the lasso
   - principal component regression and partial least squares regression.

2. Deciding how complex a model needs to be to account for the data adequately.

The purpose of this chapter is to describe some techniques for choosing the best set of predictors (and how large the estimated coefficients should be) for a multiple regression. This is a classic problem within psychology. It has received much recent attention in statistics because it is also an important area in data mining and bioinformatics (two hot areas in statistics). While in data mining and bioinformatics you may have hundreds of predictor variables, in psychology you usually have no more than about ten. Further, in data mining and bioinformatics it is often the case that you have no theory that individuates each of these variables, so you need to use some method to help find a good set of these. In psychology you usually do have some theory about the individual predictors. An exception may be education where you may have dozens (or hundreds) of questions on the typical exam, but item response methods (a form of latent variable model) are well established and cover this (Bartholomew et al., 2002; Embretson & Reise, 2000). The techniques described in this section are for exploratory analysis. If you have some particular set of theories which describes how the variables may relate, ANCOVA-type methods described in Chapter 4 are probably better suited.

The techniques described in this chapter are designed to examine only main effects. While there are procedures which also search for interactions among predictor variables, it is unlikely that many psychology data sets will require searching for interactions in an exploratory fashion so we do not cover these. It is important to realize that these techniques are only important when predictor variables are correlated among themselves (i.e., there is collinearity). If your predictor variables are uncorrelated (like when they are experimental factors) or have small correlations, then model selection is less problematic.

We look at four ways of model selection/shrinkage (more statistical details on each of these is in Hastie et al., 2001). By model selection we mean choosing a subset of predictor variables. By shrinkage we mean constraining the coefficients in such a way to increase their reliability. The first method is examining all possible sets of predictors in ordinary least square (OLS) regressions and choosing one that fits best according to some criterion. This is called *best subset regression* (bsr). The second is constraining the coefficients so that the sum of their (standardized) squared values is less than some constant. This is called *ridge regression*. All the variables are still included in the model (so there is no model selection), but the coefficients have been shrunk. Given that science is often about parsimony, it is useful to have some of the variables drop out. It turns out that if you constrain the sum of the squared absolute values of the standardized coefficients, you get variables dropping out. This is called *the lasso*, which is a catchy name because it both presents a good visual metaphor for how the coefficients are constrained and means its inventor is shown on the lasso web page in full cowboy regalia. In the last few years some extensions to the lasso have been developed, but their use is more geared towards very large numbers of predictors, as occurs with the analysis of micro-arrays in bioinformatics (and what the data mining people call support vector machines). The final technique we discuss uses principal component analysis to summarize variables. Two versions of this (PCR and PLS) are considered.

Example 6 – PTSD symptoms in fathers after child birth

- Data: From Ayers et al. (2007).
- Packages: leaps, MASS, lars, pls.
- Research question: What are the predictors of PTSD in fathers?
- Purpose: To discuss model selection in regression. To introduce several methods.

PRELIMINARY EXPLORATORY ANALYSES

The data come from Ayers and colleagues (2007) examining male and female PTSD (post traumatic stress disorder) symptoms following childbirth (a subset of the variables is used here). There are 10 predictor variables and the response variable is **PTSD**, which stands for amount of PTSD symptoms, and we are just looking at fathers. The **PTSD** variable is skewed (1.65, bootstrap 95% BCa confidence interval of (1.05, 2.37); see Chapter 2 for procedure and discussion about transforming skewed variables). Therefore, this variable was transformed: **ptsd <- log(PTSD + 1)** and the transformed variable was unskewed (0.09, 95% CI of (−0.22, 0.47)).[1] Because there were a lot of zeroes

---

[1]Box and Cox (1964) recommend examining a series of transformations of the form:

$$newX = ((oldX + s)^k - 1)/k, \text{ for } k \neq 0. \quad newX = \log(oldX + s) \text{ for } k = 0.$$

where k is the power the variable is taken to, and s is the starting value (usually used to prevent small values for the original variable becoming large negative values). When $k = 0$ it has a special form so that the transformation is continuous. The transformation used here is one of these ($k = 0$, $s = 1$). If you wanted to use the Box-Cox transformation, the **box.cox** (or **bc**) function from the **car** library (Fox, 2002, 2008) will do this: **bc(x,0,1)** is the same as **log(x+1)**. Fox (2002) describes functions for searching for values of s and k which are optimal in one sense (like being the most normally distributed).

in the original variable, there are also a lot of zeroes in this transformed variable (i.e., $\log(1) = 0$). This section is divided into four additional parts: one for each of the techniques. To load the data and transform the response variable:

```
webreg <- "http://www.sage.co.uk//wrightandlondon//"
maleptsd <- read.table(paste(webreg,"maleptsd.dat",sep=""), header=T)
attach(maleptsd)
ptsd <- log(PTSD + 1)
```

A new object with all the predictor variables is created below. This saves having to re-type all the predictor variables each time that you want to put them all in a model. **cbind** means column bind and (**rbind** means row bind; **c** will not work because it would combine all the numbers into one long variable). After doing this, if you want to refer to all the predictors you just type **preds**. If you want to refer to a subset of these, say the first four, you can type **preds[,1:4]**.

```
preds <- cbind(OVER2,OVER3,OVER5,BOND,
        POSIT,NEG,CONTR,SUP,CONS,AFF)
```

The procedures described in this chapter are necessary when the predictor variables are correlated, what is called *collinearity*. To examine the correlations we create a correlation matrix, **ptsdcorrmat**, and then print it. The option **digits=2** tells R that the minimum number of non-zero leading digits to print in any column is two. If you do not use this option, or just type **cor(preds)**, the printed correlation matrix has far more digits for each correlation than is appropriate.

```
ptsdmat <- cor(preds)
print(ptsdmat,digits=2)
```

	OVER2	OVER3	OVER5	BOND	POSIT	NEG	CONTR	SUP	CONS	AFF
OVER2	1.0000	0.4980	-0.15	-0.035	-0.22	0.528	-0.278	-0.3943	0.0088	-0.126
OVER3	0.4980	1.0000	-0.15	0.138	-0.22	0.437	-0.045	-0.3009	-0.0075	0.070
OVER5	-0.1469	-0.1541	1.00	0.189	0.54	-0.208	0.265	0.2551	0.2443	0.197
BOND	-0.0346	0.1376	0.19	1.000	0.29	-0.127	0.149	0.0853	0.2505	0.077
POSIT	-0.2193	-0.2182	0.54	0.289	1.00	-0.176	0.498	0.4676	0.2618	0.407
NEG	0.5276	0.4369	-0.21	-0.127	-0.18	1.000	-0.283	-0.2233	0.1128	-0.053
CONTR	-0.2783	-0.0454	0.26	0.149	0.50	-0.283	1.000	0.5227	0.1132	0.207
SUP	-0.3943	-0.3009	0.26	0.085	0.47	-0.223	0.523	1.0000	0.0093	0.153
CONS	0.0088	-0.0075	0.24	0.250	0.26	0.113	0.113	0.0093	1.0000	0.523
AFF	-0.1262	0.0703	0.20	0.077	0.41	-0.053	0.207	0.1532	0.5232	1.000

When **cor** creates a correlation matrix it makes it a square matrix. It has the same values on the upper triangle of the matrix as on the lower triangle.[2] So, the correlation of **OVER2** with **OVER3** is the same as **OVER3** with **OVER2** (both 0.4980).

---

[2]A matrix can be divided into 3 parts: the diagonal (which are the `1.0000`s in this matrix), the upper triangle which is the numbers above and to the right of the diagonal, and the lower triangle which is the numbers below and to the left of the diagonal.

Also, the value 1.000 is printed on the diagonal. This is because the correlation between any variable and itself will always be 1.0000. This means some of the information above is repeated (the correlations on half of the table) and some is unnecessary (the 1.0000s on the diagonal). To make the table more useful you can print more information in the table, like replacing the 1.0000 on the diagonal with the standard deviations for each variable (found by typing **sd(preds)**) and replacing one of the triangles with another measure of association. For example, Spearman's correlation is found by **cor(preds,method="spearman")**, where **spearman** is not capitalized. Alternatively, the confidence intervals or *p* values, found with the **cor.test** function for each pair of variables, could be printed. These functions are illustrated below:

```
print(sd(preds),digits=2)
```

```
OVER2 OVER3 OVER5 BOND POSIT NEG   CONTR SUP CONS AFF
3.3   3.1   1.3   3.1  11.0  11.6 14.8  5.9 11.1 3.1
```

```
spmat <- cor(preds,method="spearman")
print(spmat,digits=2)
```

```
          OVER2    OVER3    OVER5 BOND    POSIT NEG    CONTR  SUP       CONS    AFF
OVER2   1.0000   0.4980  -0.15 -0.035 -0.22  0.528 -0.278 -0.3943  0.0088 -0.126
OVER3   0.4980   1.0000  -0.15  0.138 -0.22  0.437 -0.045 -0.3009 -0.0075  0.070
OVER5  -0.1469  -0.1541   1.00  0.189  0.54 -0.208  0.265  0.2551  0.2443  0.197
BOND   -0.0346   0.1376   0.19  1.000  0.29 -0.127  0.149  0.0853  0.2505  0.077
POSIT  -0.2193  -0.2182   0.54  0.289  1.00 -0.176  0.498  0.4676  0.2618  0.407
NEG     0.5276   0.4369  -0.21 -0.127 -0.18  1.000 -0.283 -0.2233  0.1128 -0.053
CONTR  -0.2783  -0.0454   0.26  0.149  0.50 -0.283  1.000  0.5227  0.1132  0.207
SUP    -0.3943  -0.3009   0.26  0.085  0.47 -0.223  0.523  1.0000  0.0093  0.153
CONS    0.0088  -0.0075   0.24  0.250  0.26  0.113  0.113  0.0093  1.0000  0.523
AFF    -0.1262   0.0703   0.20  0.077  0.41 -0.053  0.207  0.1532  0.5232  1.000
```

```
cor.test(OVER2,OVER3)
```

```
Pearson's product-moment correlation
data: OVER2 and OVER3
t = 4.5218, df = 62, p-value = 2.824e-05
alternative hypothesis: true correlation is not equal to 0
95 percent confidence interval:
0.2873632 0.6626845
sample estimates:
  cor
0.4979958
```

These values could all be painstakingly re-typed, but luckily R allows the matrices to be added.

```
new <- diag(sd(preds)) + upper.tri(spmat)*spmat +
   lower.tri(ptsdmat)*ptsdmat
```

The **diag(sd(preds))** says to put the standard deviations on the diagonal. The **upper.tri(spmat)*spmat** says to take all values in the upper triangle of **spmat** (Spearman's ρs) and put them into the new matrix, and the **lower.tri(ptsdmat)* ptsdmat** does the same for the lower triangle and Pearson's r. Here is the new matrix:

```
print(new,digits=2)
```

	OVER2	OVER3	OVER5	BOND	POSIT	NEG	CONTR	SUP	CONS	AFF
OVER2	3.3389	0.5420	-0.25	0.036	-0.22	0.479	-0.312	-0.5444	0.0618	-0.0261
OVER3	0.4980	3.1344	-0.23	0.139	-0.14	0.414	-0.057	-0.3536	0.1098	0.0480
OVER5	-0.1469	-0.1541	1.34	0.261	0.43	-0.160	0.230	0.3568	0.1648	0.0953
BOND	-0.0346	0.1376	0.19	3.065	0.29	-0.096	0.155	0.0608	0.3159	0.1061
POSIT	-0.2193	-0.2182	0.54	0.289	11.02	-0.229	0.537	0.4818	0.2592	0.3395
NEG	0.5276	0.4369	-0.21	-0.127	-0.18	11.601	-0.362	-0.2958	0.1053	-0.0016
CONTR	-0.2783	-0.0454	0.26	0.149	0.50	-0.283	14.841	0.5231	0.1484	0.1119
SUP	-0.3943	-0.3009	0.26	0.085	0.47	-0.223	0.523	5.8654	-0.0053	0.1022
CONS	0.0088	-0.0075	0.24	0.250	0.26	0.113	0.113	0.0093	11.1466	0.4478
AFF	-0.1262	0.0703	0.20	0.077	0.41	-0.053	0.207	0.1532	0.5232	3.0676

To make the table more useful, correlations above .3 in magnitude have been underlined. From the number of underlines it is clear that many of the predictor variables are correlated, and therefore much care is needed when interpreting the individual regression coefficients. The relationships between these pairs of variables should also be examined graphically. There is a function, **scatterplot.matrix** (or **spm** for short), in the library **car** (Fox, 2008). You will need to install this package before loading it. Histograms are printed on the diagonal with a smooth line to show the distribution (Figure 5.1).

```
library(car)
spm(preds)
```

BEST SUBSET REGRESSION (bsr)

The leaps package

With 10 predictor variables, the 'best' estimate, in some sense, for how they relate to PTSD is the standard OLS regression with all the predictor variables. This is the model that has the largest R^2 and the smallest residual sums of squares. This can be calculated with the **lm** function in R as:

```
allvars <- lm(ptsd ~ preds)
summary(allvars)

Call:
lm(formula = ptsd ~ preds)

Residuals:
     Min       1Q    Median       3Q      Max
-2.06245 -0.92718  0.08414  0.80786  2.80523
```

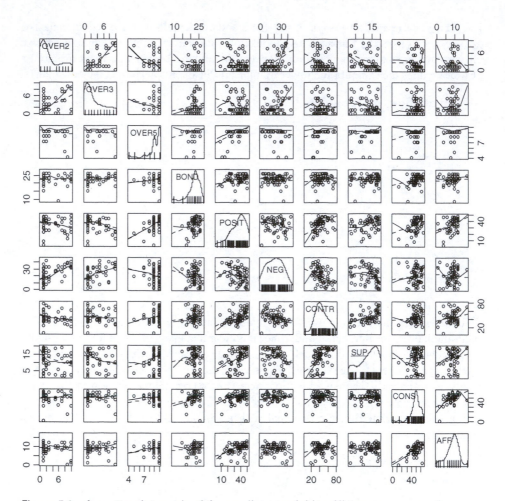

Figure 5.1   A scatterplot matrix of the predictor variables. Histograms are on the diagonal

```
Coefficients:
              Estimate Std. Error t value Pr(>|t|)
(Intercept)  1.737e+00  1.606e+00   1.081   0.2846
predsOVER2  -7.143e-02  5.793e-02  -1.233   0.2230
predsOVER3   1.191e-01  6.449e-02   1.848   0.0703 .
predsOVER5  -1.235e-06  1.332e-06  -0.927   0.3581
predsBOND   -8.802e-02  5.512e-02  -1.597   0.1162
predsPOSIT   2.490e-02  2.051e-02   1.214   0.2301
predsNEG     3.593e-02  1.679e-02   2.140   0.0370 *
predsCONTR  -8.588e-03  1.316e-02  -0.653   0.5167
predsSUP     4.130e-02  3.274e-02   1.261   0.2127
predsCONS    1.473e-02  1.717e-02   0.857   0.3951
predsAFF     3.635e-02  6.415e-02   0.567   0.5734
---
Signif. codes:  0 '***' 0.001 '**' 0.01 '*' 0.05 '.' 0.1 ' ' 1
```

```
Residual standard error: 1.155 on 53 degrees of freedom
Multiple R-Squared: 0.3047,      Adjusted R-squared: 0.1736
F-statistic: 2.323 on 10 and 53 DF,  p-value: 0.02368
```

This shows a multiple R^2 of .30. The model with all these variables is a complex. None of the coefficient estimates can be easily interpreted because each is conditional on nine others. Thus, you cannot say the **NEG** is positively related to **ptsd** from this output (you would use **cor.test(NEG,ptsd)** to say this). From the multiple regression output you have to say it as: '**NEG** is positively related to **ptsd** after partialling out **OVER2**, **OVER3**, **OVER5**, **BOND**, **POSIT**, **CONTR**, **SUP**, **CONS**, and **AFF**.' It is doubtful that any human could really understand what this would mean for any theory, particularly as many of the variables are correlated among themselves. A goal of science is to simplify models like this. One of the standard stepwise methods (backwards stepwise) is to take one variable away (it would be **AFF** since it is the least significant) and then re-examine the model. The main problem with this automated approach is that it does this automatically based on something as meaningless as the p value. In the simplest circumstances p values have a difficult meaning (Cohen, 1990, 1994; Dienes, 2008), but here, where they are based on a relationship after partialling out nine other variables, the p values have essentially no scientific value. Why take **AFF** out rather than **CONTR** or **OVER5** or any of them? This is the main reason methodologists stress avoiding these automated methods, but another reason is that these methods are not guaranteed to find the best subset, and there are methods which address this deficiency.

One alternative is to say, for all of the models with k predictor variables, which is the one with the best fit. When the number of variables is very large the number of possible models to check is 2^k where k is the number of variables. If you have 40 variables, this is (using **2^40** in R) approximately 10^{12}, or 10 with 11 more zeroes, which is far too many to deal with. Even if the computer could solve a thousand models every second, it would take over thirty years. And, if you had 70 variables, the necessary time is longer than current estimates of the age of the universe.[3] Even smaller numbers of variables can create problems, so statisticians have created clever algorithms that search for the best fits. Some packages, like SAS, have built-in procedures for *best subset regression* (*bsr*). Others, like SPSS, require additional instructions (http://distdell4.ad.stat.tamu.edu/spss_1/allpos1.html; syntax for doing this is: http://www.spsstools.net/Syntax/RegressionRepeatedMeasure/DoAll-SubsetsRegressions.txt).

Within R, the **leaps** package (Lumley, 2006) is recommended for this procedure. This will need to be installed (see Chapter 1) and loaded. There are many different criteria for estimating the fit of the model. Cross-validation is often used for exploring the fit of different models. This involves fitting a model to a subset of the data, seeing how the model fits the remaining data, and then doing this for several subsets. This is discussed later in this chapter. A simpler method, and much more common in psychology, is to summarize the fit of the model to the entire data set with a single statistic. The statistics available in **leaps** include Mallow's Cp, BIC, AIC, R^2, and adjusted R^2.

---

[3]A recent estimate for the age of the universe from NASA is 1.37×10^{10} years (http://map.gsfc.nasa.gov/m_mm/mr_age.html, accessed March 17, 2007), though this requires some assumptions, and some scientific models do suggest it may be infinite. Some clubs, cults, and religions provide much smaller estimates and some say it is infinite.

R^2 and adjusted R^2 are popular in psychology, so we will use these. The first two commands install and load **leaps**. You may need to choose a mirror site from which to download the package.

```
install.packages("leaps")
library(leaps)
```

The next command runs a series of regressions which searches for the best fitting models for each number of predictor variables. So, with 10 potential predictors there can be anywhere between 0 and 10 predictors in a model. The **summary** command shows which variables are to be included at each step. So, if one variable is to be used, it is **NEG**, if two are used they are **NEG** and **AFF**. If eight variables are to be used then they are all the variables except **CONTR** and **AFF**. So **AFF** is no longer in the model, despite being in before. The graph produced is difficult to read if the number of variables gets larger and it is not visually appealing as is. At least all the **"**s should be removed if printing it for a paper or an assignment.

```
x1 <- regsubsets(preds,ptsd)
summary(x1)
```

```
Subset selection object
10 Variables  (and intercept)
      Forced in Forced out
OVER2     FALSE      FALSE
OVER3     FALSE      FALSE
OVER5     FALSE      FALSE
BOND      FALSE      FALSE
POSIT     FALSE      FALSE
NEG       FALSE      FALSE
CONTR     FALSE      FALSE
SUP       FALSE      FALSE
CONS      FALSE      FALSE
AFF       FALSE      FALSE
1 subsets of each size up to 8
Selection Algorithm: exhaustive
          OVER2 OVER3 OVER5 BOND POSIT NEG CONTR SUP CONS AFF
1  ( 1 )  " "   " "   " "   " "  " "   "*" " "   " " " "  " "
2  ( 1 )  " "   " "   " "   " "  " "   "*" " "   " " " "  "*"
3  ( 1 )  " "   " "   " "   " "  " "   "*" " "   "*" " "  "*"
4  ( 1 )  " "   " "   " "   "*"  " "   "*" " "   "*" " "  "*"
5  ( 1 )  " "   "*"   " "   "*"  " "   "*" " "   "*" " "  "*"
6  ( 1 )  "*"   "*"   " "   "*"  " "   "*" " "   "*" " "  "*"
7  ( 1 )  "*"   "*"   " "   "*"  "*"   "*" " "   "*" " "  "*"
8  ( 1 )  "*"   "*"   "*"   "*"  "*"   "*" " "   "*" "*"  " "
```

The package **leaps** has some graph capabilities which we will now look at. We set it up so that the graphs are in a 2 × 2 grid format with the **par(mfrow=c(2,2))** command. The **leaps** graphs are in the top two panels of Figure 5.2. The bottom panels take information from the object created by the **regsubsets** function and present the information in a more visually appealing manner. The " * " plot printed above has no scale other than the number of variables in the model. This is fine if all you want to see is if model x has a better or worse fit than model y, if they have the same number of predictors, but it does not tell you how much better or allow you to compare between models with different numbers of predictors. To do this you need to define a measure of fit. The **scale=c("r2")** and **scale=c("adjr2")** are used with **plot** to show whether R^2 or adjusted R^2 should be used. The **plot** function treats these objects in special ways (i.e., different from **lm.objects**).

```
par(mfrow=c(2,2))
plot(x1,scale=c("r2"))
title("Default for leaps")
plot(x1,scale=c("adjr2"))
title("Default for leaps")
```

More visually appealing graphs can be constructed by running the **leaps** function and plotting parts of the resulting objects. The following commands create two objects, **x2** and **x3**, which show the best fitting (**nbest=1** means the command just stores the best model) regression model for each number of predictor variables. **leaps** allows different measures to determine how to measure fit. For **x2** R^2 is used and for **x3** adjusted R^2 is used. The number of variables in the model, including the intercept, can be found with **x2$size**.

```
x2 <- leaps(preds,ptsd,nbest=1, method="r2")
x3 <- leaps(preds,ptsd,nbest=1, method="adjr2")
plot(x2$size-1,x2$r2,xlab="Number of predictors",
   ylab=expression(R^2))
lines(spline(x2$size-1,x2$r2))
plot(x3$size-1,x3$adjr2, xlab="Number of predictors",
   ylab=expression(adj.R^2))
lines(spline(x3$size-1,x3$adjr2))
par(mfrow=c(1,1))
```

Figure 5.2 provides information to help choose the number of predictors to have in your model. The default plots for **leaps** seem difficult to read. Below these are the plots for R^2 and adjusted R^2 which are easier to read. They show the fit statistics extracted from the **leaps** objects for the best models for each number of predictor variables. **x2** is a **leaps** object and because it was made with **method="r2"** the R^2 values can be extracted with **x2$r2**. Similarly for **adjr2** with **x3**. We have used **x2$size-1** and **x3$size-1** for the number of predictors since **leaps** counts the intercept as one of the variables. Notice that the R^2 values continue to increase each time you add a variable, but the adjusted R^2 value hits a peak. This is why the adjusted value is often used to decide between models; it goes down when the predictor variable added has no predictive value. Statisticians say that the statistic, adjusted R^2, penalizes models with lots of predictor variables.

Figure 5.3 shows the lower right-hand graph from the 2×2 grid of Figure 5.2 but in a more useful way. Notice that the adjusted R^2 value is higher with three predictors than with four predictors. If you were using a forward stepwise method to search for models and stopped the search if none of the variables not included in the model increased the

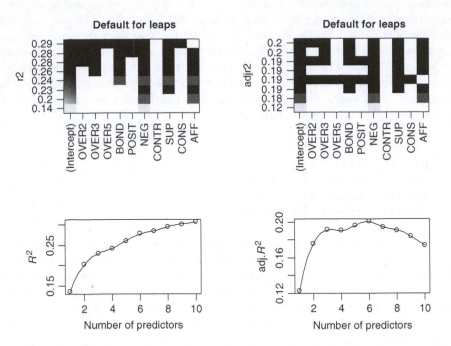

Figure 5.2 The default plots for the **leaps** function and plots showing the amount of predictor variables and the R^2 and adjusted R^2

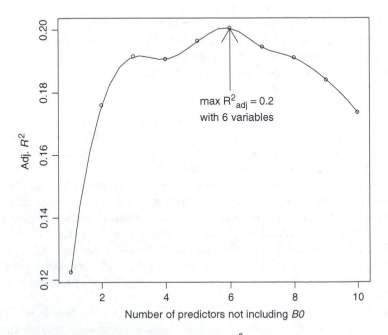

Figure 5.3 A plot showing the maximum adjusted R^2

adjusted R^2, you would stop at the model with just **NEG**, **SUP**, and **AFF**. We have used **expression** and **paste** in the **text** functions below. These allow you to put in mathematical expressions and variable values. The **pos** in the **text** command tells R where the text should be placed in relation to the *x*, *y* coordinates given at the start of the command (1 is below, 2 is to the left, 3 is above, and 4 is to the right).

```
plot(x3$size-1,x3$adjr2,xlab="# of predictors not
    including B0", ylab="Adj. R-squared")
lines(spline(x3$size-1,x3$adjr2))
m2 <- max(x3$adjr2)
m1 <- which.max(x3$adjr2)
arrows(m1,m2-.02,m1,m2)
text(m1,m2-.023,expression(paste("max ",
    r[adj]^2 == 0.2)))
text(m1,m2-.031,paste("with",format(m1,digits=1),
    "variables"),pos=3)
```

Once you have decided how many predictors you want, you need to find out what they are, and then evaluate the regression model with them. The **which** and **which.max** commands allow you to find which variables are included in the model that has the maximum adjusted R^2. It is the 1, 2, 4, 6, 8 and A variables (when the function runs out of the digits 1–9 it uses letters), which are: **OVER2**, **OVER4**, **BOND**, **NEG**, **SUP**, and **AFF**.

```
x3$which[which.max(x3$adjr2),]
```

1	2	3	4	5	6	7	8	9	A
TRUE	TRUE	FALSE	TRUE	FALSE	TRUE	FALSE	TRUE	FALSE	TRUE

```
labels(preds[1,])
```

```
[1] "OVER2" "OVER3" "OVER5" "BOND"  "POSIT" "NEG"   "CONTR" "SUP"
    "CONS"  "AFF"
```

The **lm** model of these is:[4]

```
summary(lm(ptsd ~ preds[,c(1,2,4,6,8,10)]))
```

```
Call:
lm(formula = ptsd ~ preds[, c(1, 2, 4, 6, 8, 10)])

Residuals:
    Min       1Q   Median       3Q      Max
-2.01516 -0.96366 -0.03749  0.87906  2.82240
```

---

[4]All of these steps can be combined with:
```
summary(lm(ptsd~preds[,x3$which[which.max(x3$adjr2),]]))
```

```
Coefficients:
                                  Estimate Std. Error t value Pr(>|t|)
(Intercept)                        0.70057    1.24076   0.565  0.57454
preds[, c(1,2,4,6,8,10)]OVER2     -0.06445    0.05650  -1.141  0.25873
preds[, c(1,2,4,6,8,10)]OVER3      0.08485    0.05692   1.491  0.14156
preds[, c(1,2,4,6,8,10)]BOND      -0.05972    0.04857  -1.230  0.22389
preds[, c(1,2,4,6,8,10)]NEG        0.04383    0.01520   2.883  0.00555 **
preds[, c(1,2,4,6,8,10)]SUP        0.03926    0.02719   1.444  0.15420
preds[, c(1,2,4,6,8,10)]AFF        0.08496    0.04810   1.766  0.08268 .
---
Signif. codes:  0 '***' 0.001 '**' 0.01 '*' 0.05 '.' 0.1 ' ' 1

Residual standard error: 1.136 on 57 degrees of freedom
Multiple R-Squared: 0.2767,     Adjusted R-squared: 0.2005
F-statistic: 3.633 on 6 and 57 DF,   p-value: 0.004025
```

Notice that most of the p values are above .05. The adjusted R^2 is for the model, not the individual coefficients. Most of the stepwise procedures would continue removing terms. The decision about how many terms to include should be based on how important parsimony is for your particular application. We believe parsimony is usually very important in psychology, so that researchers should usually opt for a simpler model. While adjusted R^2 penalizes complex models, other adjustments have greater penalties for complex models. The pragmatic approach to statistics, however, stresses that decisions and conclusions should not be based on any single statistic and that the researcher should look at various indices and describe the range of possible conclusions which their data may suggest.

RIDGE REGRESSION

The *bsr* (best subset regression) method described above either includes a variable or not, and often the choice of whether to include a variable is based on only a minute difference in fit. Efron et al. (2004: 409) describe this as 'overly greedy, impulsively eliminating covariates which are correlated with' other covariates. One alternative is a more smooth transition where the sizes of the regression coefficients are constrained. This is called *ridge regression* and it can be solved with a form of least squares regression. Therefore, it is a technique that has been used for several decades and can be calculated using many of the general statistics packages (for example, SPSS).

In R there are a few functions for ridge regression. One of the simplest is **lm.ridge** in the **MASS** library (Venables & Ripley, 2002), so we will use that (see also Halvorsen, 2007). It takes the sum of the squared standardized regression estimated coefficients and constrains them to be only as large as some value k:

$$\sum \hat{\beta}j^2 \leq k$$

The value k is one of several measures of the amount of shrinkage. The function **lm.ridge** plots the size of the individual coefficients with the amount of shrinkage, but uses λ (lambda) (which is used in the computation of the ridge regression). As λ goes up the amount of shrinkage goes up, and *k* goes down (Hastie et al., 2001: 59). Here is the R code for Figure 5.4:

```
library(MASS)
lm.ridge(ptsd~preds,lambda=seq(0,100,by=1))-> x
plot(x)
title("Ridge Regression")
abline(h=0)
abline(v=50,lty=3)
```

Usually you need to use trial and error to decide the range of the λs to be tested. Here, `seq(0,100,by=1)` means going from 0 to 100 in steps of 1 (even steps of 10 or 20 produce smooth enough curves, but the computer is fast enough to have 100 steps). Note that `->` is used to assign the `lm` object to `x` rather than the other way around.

HELPFUL HINT: USING THE UP ARROW

We used `->` instead of `<-` when running our ridge regression. This is acceptable in R, and is actually easier in this particular situation. Conducting many types of analyses requires some trial and error. Here we ran several ridge regressions with different step sequences. We did not save any of them, but just looked at their output. Rather than re-typing the command each time we used the up arrow on the keyboard. This is very helpful (and pressing the up arrow several times gives earlier commands). Once we were happy with the regression, we typed `->x` at the end to save it.

There are complex ways to compare the fit of models like this, but for the present purpose just choose a λ (lambda) for where the coefficients seem relatively stable. The value lambda = 50 seems about right on Figure 5.4 and we added a vertical line at this value with the **abline** function. Here are the coefficient values for lambda = 50:

```
x$coef[,50]
```

```
   predsOVER2    predsOVER3    predsOVER5     predsBOND    predsPOSIT
 -0.041302391   0.151376893  -0.065759222  -0.116512457   0.088696632
    predsNEG     predsCONTR      predsSUP     predsCONS      predsAFF
  0.242561482  -0.008260764   0.111944432   0.081908793   0.126166511
```

We are not going to spend much time on ridge regression because we do not think it is of much use. Historically it is important because it could be solved relatively easily and so has been available for decades. We presented it as a way to introduce a better procedure called the *lasso*. A problem with ridge regression is that all of the variables are still included in the model. Given that it is better to have simpler models, it would have been nice if some of these coefficients had dropped out. This occurs with the next procedure, the lasso.

Ridge regression

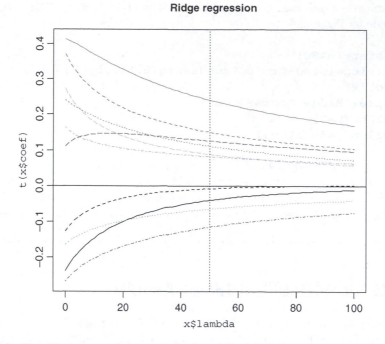

Figure 5.4   The ridge regression output for Ayers et al. (2007)

THE LASSO

The lasso works by constraining the sum of the absolute values of standardized estimated coefficients to some constant, k. In math:

$$\sum \left| \hat{\beta}j \right| \leq k$$

While the difference between this and what is done with ridge regression appears slight, there are two important consequences. First, ridge regression is computationally simpler than the lasso so standard least squares techniques can be used to estimate the coefficients. The lasso is more difficult computationally. Second, in ridge regression while the individual coefficients shrink and sometimes approach zero, they seldom reach zero so they are not excluded from the model. With the lasso the coefficients reach zero and therefore predictor variables do drop out. This means that the lasso leads to a more parsimonious model than ridge regression. In technical terms, ridge regression is a method of shrinkage, not model selection, while lasso does both.

The computational difficulty with the lasso has been solved. Efron and colleagues (2004) have developed an algorithm called least angle regression (lars) that calculates the lasso solution in about the same time as least squares regression. Therefore, the lasso should regularly be used instead of ridge regression (although other techniques exist which may be better suited for particular situations; for example, see Tibshirani et al., 2005). If k (the shrinkage value in the formula above) is chosen to be too small then the model may not capture important characteristics of the data. If k is chosen to be too large then the model may over-fit the data in the sample, providing an inaccurate representation for the population. As described above cross-validation and measures of fit like adjusted R^2 are often used to assess the fit and to decide how much to constrain the size of the coefficients. The **lars** package (Hastie & Efron, 2007) allows both the computation of lasso coefficient estimates and cross-validation to help the researcher decide the appropriate amount of shrinkage. As stated above, it is important that researchers do not rely too much on any single statistic to guide their conclusions, and again here the researcher must consider the importance of parsimony for the particular application.

lars can be downloaded in the usual way from CRAN, but it is worth looking at Hastie's web page (www-stat.stanford.edu/~hastie/Papers) for lots of other information that he has made available.

```
install.packages("lars")
library("lars")
```

To illustrate the package we will go through the same example as above (Ayers et al., 2007).

```
lasso1 <- lars(preds,ptsd)
plot(lasso1)
```

and Figure 5.5 is produced.

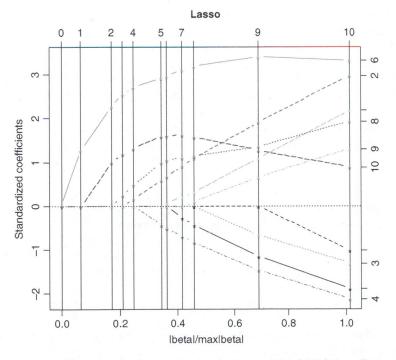

Figure 5.5   The graph of the lasso solution for the Ayers et al. (2007) data. The vertical lines show when a variable has been eliminated from the model

The scales are different than on the ridge regression graph, and the lasso graph starts at constraining the coefficients to zero and then moves towards ordinary least squares, but the basic form of the graph is the same. An advantage of the lasso over ridge regression is that variables actually drop out; their coefficients go to zero. Vertical lines are placed each time a variable drops out, and in fact, these are the only points that are actually shown in the default lasso graph (this can be changed). The variable numbers are shown to the right of the graph.

The function, **lars**, uses least angle regression to calculate the entire sequence of lasso coefficients. Lasso is the default. Here, the solution is stored in **lasso1**, which is an object, like the **lm** objects, so it can be used in other functions. In Figure 5.5 the y-axis is the standardized coefficients and the x-axis is labeled '|beta|/max|beta|' which is for: $\Sigma|\beta s|/max\Sigma|\beta s|$. Both of these labels deserve further explanation.

When using the lasso, it makes sense to remove the constant, and usually to standardize all the variables so that the shrinkage does not penalize some coefficients more simply because of their scale (this is true also with ridge regression). The **lars** package does this automatically (as does the ridge regression function discussed). The β values are stored in **lasso1$beta**. The graph uses the transformed values which can be found with the following command: **scale(lasso1$beta,FALSE,1/lasso1$normx)**. No one other than the functions' authors would be expected to know this level of information, but that is why they produced help files (**help(lars)**) and have a manual.

The '|beta|/max|beta|' on the x-axis ranges from 0 (sum of the $|\beta s|$ being zero) to 1 (no shrinkage, the OLS unbiased estimates). Rather than using k, which would depend on the scale of the variables, this amount has been transformed so that it is comparable across problems. This graph has other useful features. The curves produced by R for each covariate are in different colors on the screen so can be differentiated. The vertical lines show each time a covariate is added to the model (or eliminated from the model, depending on whether you look at it from left-to-right or right-to-left). These are referred to as steps in the output. On the far right the covariate numbers are shown. With several covariates not all of them are printed, but the user can figure out which line is for which covariate by looking at the values.

Normally methods like cross validation are used to decide how much shrinkage should be used, but for present purposes lets say the point of shrinkage where there are still six predictor variables looks good. Since the package includes the intercept here, let **s=7**, and write:

```
predict(lasso1,s=7,mode="step",
  type="coefficient")$coefficients
```

OVER2	OVER3	OVER5	BOND	POSIT
0.0000000000	0.0270736438	0.0000000000	-0.0209091602	0.0007274102
NEG	CONTR	SUP	CONS	AFF
0.0320769257	0.0000000000	0.0224833783	0.0000000000	0.0660251905

These coefficients can then be used. Another possibility is just using two predictor variables (let **s=3**).

There is a cross-validation procedure built within **lars** called **cv.lars**. Here we just run the default (usually you vary the number of folds with the option **K=**). With the

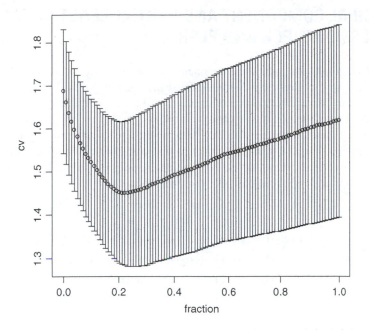

Figure 5.6   The graphical output from **cv.lars** for the Ayers et al. (2007) data

small data set used here for illustration the default, **K=10**, will do. The first command runs the procedure and produces Figure 5.6. The second calculates which value has the lowest cv score, and then finds the fraction associated with that, and stores the result in **frac**. The coefficients for this fraction are then found using the **predict** function but telling it that **mode="fraction"**. Notice that only four predictor variables are included.

```
lassocv <- cv.lars(preds,ptsd)
frac <- lassocv$fraction[which.min(lassocv$cv)]
predict(lasso1,s=frac,mode="fraction",
    type="coefficient")$coefficients
```

```
     OVER2          OVER3          OVER5          BOND           POSIT
0.000000000  0.002221251  0.000000000  0.000000000  0.000000000
       NEG          CONTR            SUP           CONS            AFF
0.028186329  0.000000000  0.006975196  0.000000000  0.050473978
```

There is a big conceptual hurdle in adopting this approach. The ordinary least squares estimates, from the **lm** function, are unbiased, but they have higher variance than other procedures. Hastie et al. (2001) discuss in detail the importance of weighting both bias and variance when choosing a statistic. Efron et al. (2004) describe a hybrid lars/OLS procedure, where **lars** is used to decide which covariates to include and then the OLS coefficients are reported. This has the advantage that the output will be more familiar to most psychologists. This is the procedure that Ayers et al. (2007) reported in their paper, but it falls foul of Efron et al.'s (2004: 409) 'overly greedy' criticism mentioned earlier.

PRINCIPAL COMPONENT AND PARTIAL LEAST SQUARES REGRESSIONS (PCR AND PLSR)

The final techniques use principal component analysis (PCA) to create predictor variables and use these in a multiple regression. Because the components will be uncorrelated, the problems of multicollinearity are not applicable. We are assuming that you are familiar with PCA. It is a data reduction technique which begins by finding a linear combination of the variables with the largest variance. It then finds the linear combination of variables that has the next largest variance subject to this combination being uncorrelated with the first. Subsequent components are found, again, subject to them being uncorrelated with the previous components. So, the four components from a PCA for four X variables would be:

$$Comp1_i = a_{11}x1_i + a_{21}x2_i + a_{31}x3_i + a_{41}x4_i$$
$$Comp2_i = a_{12}x1_i + a_{22}x2_i + a_{32}x3_i + a_{42}x4_i$$
$$Comp3_i = a_{13}x1_i + a_{23}x2_i + a_{33}x3_i + a_{43}x4_i$$
$$Comp4_i = a_{14}x1_i + a_{24}x2_i + a_{34}x3_i + a_{44}x4_i$$

where the a values are called loadings, which are estimated. For interpretability, it is hoped that many of these loadings are near zero so can be ignored. Sometimes all of the loadings for later components are small enough that these components can be ignored without much loss of information. Because of this PCA is used as a data reduction/simplification technique.

PCA is often likened to exploratory factor analysis (EFA). It has some superficial similarities with EFA, for example, they both rely on correlations among the X variables to yield useful results and that correlated variables tend to hang together in both the components of PCA and the factors of EFA. PCA is preferred by most statisticians. The difficulty many people have with EFA is that hypothesizing these latent variables and measuring them, often in the same step, is a dubious scientific approach and should only be done with great caution. Bartholomew et al. (2002) provide a good treatment of each and compare them.

It is possible to use the R function **princomp** for PCA and then enter the resulting components into a linear regression (**lm**). Alternatively, the **pls** library (Wehrens & Mevik, 2007) allows these two steps to be done with a single function for principal component regression (**pcr**), and also provides useful options and an alternative. The alternative is *partial least squares regression* (**plsr**) which combines the procedure into a single step. Rather than finding the linear combination of X variables with the largest variance, **plsr** uses information about both the X variables and their associations with the response variable. The choice between these alternatives depends on your particular goals. If you want to reduce the dimensionality of a large number of variables into a small number of components, and then see how these components predict a response variable, use **pcr**. If you want to see how well a combination of a set of variables can predict a response variable, use **plsr**. **plsr** will produce a better prediction of the response variable, because that is what it is designed to do, but it will often be more difficult to interpret. The examples below show how these methods yield different results.

Several algorithms are offered for the **plsr** approach, but they produce the same solutions when there is only a single Y variable. We show both the **pcr** and the **plsr** approaches for the Ayers et al. (2007) data. We use the **pls** library which you will need to install and load.[5]

```
install.packages("pls")
library("pls")
```

This first set of code runs **pcr** and produces Figure 5.7 which can be used to help decide how many components you need (like how you would use a scree plot).[6] The values are the cumulative percentage variance accounted for. It is the non-cumulative ones that go in a normal scree plot. The values for **ptsd** is how much of the **ptsd** variance is accounted for by the components. The first accounts for nothing (well, 0.004%). The second accounts for 13%, which is a relatively large amount. If using all 10 components, 30.47% is accounted for. This is the same as the multiple regression using all the variables. This makes sense since it is the same information.

```
pcr1 <- pcr(ptsd~preds,ncomp=10)
plot(1:10,summary(pcr1)[1,],ylab="Cumulative
    variance accounted for",xlab="Number of components",
    ylim=c(0,100),pch=19,cex.lab=1.3)
```

```
Data:     X dimension: 64 10
          Y dimension: 64 1
Fit method: svdpc
Number of components considered: 10
TRAINING: % variance explained
            1 comps   2 comps   3 comps   4 comps   5 comps   6 comps   7 comps
X       45.068953     67.21     82.13     92.42     95.83     97.30     98.43
ptsd     0.004225     13.07     17.16     17.91     19.82     20.07     25.39
            8 comps   9 comps   10 comps
X          99.36     99.83     100.00
ptsd       27.11     29.24      30.47
```

```
lines(1:10,summary(pcr1)[1,],lwd=1.5)
points(1:10,summary(pcr1)[2,],pch=19)
lines(1:10,summary(pcr1)[2,],lwd=1.5)
```

---

[5]Both of these functions, **pcr** and **plsr**, can be accessed as different methods from a function called **mvr** (for multivariate regression), but we will keep with **pcr** and **plsr** so that the function name indicates what is being done. The function name **mvr** indicates that it can be used, generally, for multiple response variables.
[6]An odd thing occurs with the **pcr** function. Every time you use **summary(pcr.object)** or **summary(plsr1)**, which we use below, the computer echoes the summary output to the screen. It is likely the **invisible** function, used in the **mediator** function of Chapter 4, could take care of this. Anyway, this is a minor nuisance rather than a problem but worth mentioning because if you run these functions you may get more output than is printed here in this book. Throughout the rest of this chapter, when R echoes the summary, we will not print it in order to save a little paper.

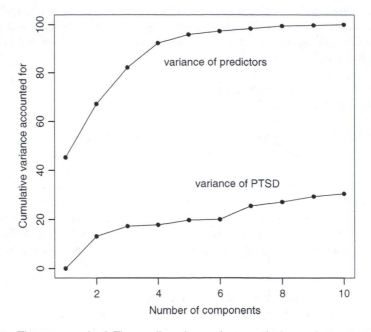

Figure 5.7 The **pcr** method. The top line shows the cumulative variance accounted for by each of the components, the same as you would find with a PCA. The bottom line shows how much these components account for the **ptsd** variable

```
text(4,85,"variance of predictors",pos=4,cex=1.2)
text(5,35,"variance of PTSD",pos=4,cex=1.2)
```

Like with other principal components software, R will produce the loadings without showing some that are low so that it is easier to see which variables load onto which components. If you type **pcr1$projection** you get all the loadings. Because of Figure 5.7, you would probably focus just on the first two components because none of the others account for much variation in **ptsd**. Here are these loadings.

pcr1$loadings

```
Loadings:
        Comp 1 Comp 2 Comp 3 Comp 4 Comp 5 Comp 6 Comp 7 Comp 8 Comp 9 Comp 10
OVER2                             -0.241  0.329 -0.245  0.799 -0.336
OVER3                 -0.122      -0.237  0.520 -0.445 -0.176  0.647
OVER5                                                         0.107  0.990
BOND                             -0.103  0.719  0.461 -0.335 -0.373
POSIT  -0.459 -0.245  0.128 -0.809 -0.205
NEG     0.306 -0.618 -0.685 -0.124                     -0.108
CONTR  -0.786        -0.450  0.395 -0.123
SUP    -0.205               -0.139  0.896  0.298 -0.146  0.137
CONS   -0.148 -0.729  0.520  0.385
AFF           -0.106                            -0.703 -0.421 -0.543
```

	Comp 1	Comp 2	Comp 3	Comp 4	Comp 5	Comp 6	Comp 7	Comp 8
SS loadings	1.0	1.0	1.0	1.0	1.0	1.0	1.0	1.0
Proportion Var	0.1	0.1	0.1	0.1	0.1	0.1	0.1	0.1
Cumulative Var	0.1	0.2	0.3	0.4	0.5	0.6	0.7	0.8

	Comp 9	Comp 10
SS loadings	1.0	1.0
Proportion Var	0.1	0.1
Cumulative Var	0.9	1.0

The following shows that the solution using this approach to **pcr** is the same as running **princomp** (for PCA) and then **lm** on the components. The principal component analysis is done and stored in **pca1**, and the scores for the first two components are used to predict **ptsd**. The multiple R^2 shows that 13.07% of the variance is accounted for.

```
pca1 <- princomp(preds)
pcrreg <- lm(ptsd ~ pca1$scores[,1] + pca1$scores[,2])
summary(pcrreg)


Call:
lm(formula = ptsd ~ pca1$scores[, 1] + pca1$scores[, 2])


Residuals:
    Min       1Q    Median       3Q      Max
-2.28493 -1.03742  0.07687  0.84556  2.46798


Coefficients:
                     Estimate Std. Error t value Pr(>|t|)
(Intercept)         1.5924224  0.1505042  10.581 1.97e-15 ***
pca1$scores[, 1]   -0.0004731  0.0086882  -0.054   0.9568
pca1$scores[, 2]    0.0375420  0.0123959   3.029   0.0036 **
---
Signif. codes:  0 '***' 0.001 '**' 0.01 '*' 0.05 '.' 0.1 ' ' 1


Residual standard error: 1.204 on 61 degrees of freedom
Multiple R-Squared: 0.1307,      Adjusted R-squared: 0.1022
F-statistic: 4.588 on 2 and 61 DF,   p-value: 0.01393
```

The partial least square regression (**plsr**) procedure is now shown. The syntax works the same way so the same type of graph is produced in Figure 5.8. Because this procedure tries to account for variation within the response variable, the first component does this while still trying to combine the predictor variables.

```
plsr1 <- plsr(ptsd~preds,ncomp=10)
plot(1:10,summary(plsr1)[1,],
  ylab="Cumulative variance accounted for",
  xlab="Number of components",ylim=c(0,100),
  pch=19,cex.lab=1.3)
lines(1:10,summary(plsr1)[1,],lwd=1.5)
```

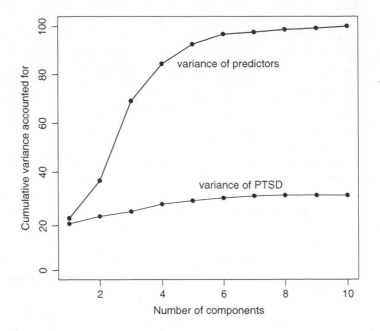

Figure 5.8   The figure for **plsr**. The top line shows the cumulative amount of variance in the predictors accounted for by the components. The bottom line shows the amount accounted for of the variables **ptsd**. One component seems fine

```
points(1:10,summary(plsr1)[2,],pch=19)
lines(1:10,summary(plsr1)[2,],lwd=1.5)
text(4.3,85,"variance of predictors",
  pos=4,cex=1.2)
text(5,35,"variance of PTSD",pos=4,cex=1.2)
```

The numbers below are those used in Figure 5.8. Notice that there is a much higher amount of **ptsd** variance accounted for than with **pcr**. It is still less than the amount from OLS regression including everything. **plsr** can be seen as a combination of these.

summary(plsr1)

```
Data:    X dimension: 64 10
         Y dimension: 64 1
Fit method: kernelpls
Number of components considered: 10
TRAINING: % variance explained
        1 comps  2 comps  3 comps  4 comps  5 comps  6 comps  7 comps
X        21.07    36.40    68.99    84.41    92.75    96.82    97.49
ptsd     18.81    21.78    23.75    26.88    28.24    29.41    30.17
        8 comps 9 comps  10 comps
X        98.53   99.20    100.00
ptsd     30.34   30.45    30.47
```

plsr1$loadings

```
Loadings:
```

	Comp 1	Comp 2	Comp 3	Comp 4	Comp 5	Comp 6	Comp 7	Comp 8	Comp 9	Comp 10
OVER2	0.109		-0.198			-0.424		0.782	-0.728	
OVER3					0.275	0.247	0.509	-0.558	0.370	-0.415
OVER5							-0.145	-0.199	-0.418	-0.248
BOND		-0.115	-0.140		-0.167		0.418	-0.970	0.407	0.283
POSIT	0.233	0.346	-0.677	0.472	-1.029	0.766		0.112		
NEG	0.839	0.210	0.429	-0.487	0.193	-0.105			-0.147	0.164
CONTR		0.796	-1.415	0.405	0.587	-0.299				
SUP		0.471	-0.132	0.366	-0.248	-0.489	0.326	-0.271	0.361	-0.283
CONS	0.522	-0.998	-0.300	0.563		-0.232	0.123			
AFF				0.219		0.212	-0.730	-0.299	0.283	0.208

	Comp 1	Comp 2	Comp 3	Comp 4	Comp 5	Comp 6	Comp 7	Comp 8
SS loadings	1.060	2.036	2.778	1.170	1.622	1.090	1.296	1.477
Proportion Var	0.106	0.204	0.278	0.117	0.162	0.109	0.130	0.148
Cumulative Var	0.106	0.310	0.587	0.704	0.867	0.976	1.105	1.253

	Comp 9	Comp 10
SS loadings	1.327	1.000
Proportion Var	0.133	0.100
Cumulative Var	1.386	1.486

This shows that the final part of the **plsr** procedure is a regression with the scores.

```
plsreg <- lm(ptsd ~ plsr1$scores[,1])
summary(plsreg)
```

```
Call:
lm(formula = ptsd ~ plsr1$scores[, 1])

Residuals:
     Min       1Q   Median       3Q      Max
-2.27623 -0.98499  0.02705  0.85712  2.33879

Coefficients:
                   Estimate Std. Error t value Pr(>|t|)
(Intercept)         1.59242    0.14428   11.04 2.87e-16 ***
plsr1$scores[, 1]   0.04752    0.01254    3.79 0.000344 ***
---
Signif. codes:  0 '***' 0.001 '**' 0.01 '*' 0.05 '.' 0.1 ' ' 1

Residual standard error: 1.154 on 62 degrees of freedom
Multiple R-Squared: 0.1881,    Adjusted R-squared: 0.175
F-statistic: 14.36 on 1 and 62 DF,  p-value: 0.0003439
```

PRINCIPAL COMPONENTS VERSUS FACTOR ANALYSIS

Several people have asked why principal component analysis is used rather than exploratory factor analysis. They ask if this is because of the generally negative view that many statisticians have of factor analysis. The answer is complex. We have nothing against latent variable models (and mentioned them earlier with item response models and use them in some of our own research). You could use factor analysis, save the factors, and use them in a regression (a latent variable analog to PCR), or use structural equation modeling which would use information from both the predictor and response variables (a latent variable analog to PLSR). The difficulty, as with exploratory factor analysis, is trying to get a handle on what these latent variables are. Some people write about factor analysis as if these factors are 'true scores' of some imagined attribute, and in fact that is what is assumed in the creation of them. However, given that in practice the scores are just a combination of the observed variables then it is often easier to use techniques designed to combine these observed variables. So, the warning on using latent variable approaches is that you may come up against a philosopher who will argue with you, and that is never pretty.

SUMMARY AND WHICH TECHNIQUE TO USE

The stepwise procedures used in much psychology are frowned upon by methodologists. The main reason methodologists dislike them is because they take the decision making away from the researcher, who is the one who understands the meaning of the variables and the relevant theories. These model selection and regularization procedures are almost exclusively data driven. There are two situations where this data driven approach is useful. The first is when you have no theory that takes into account the individual predictor variables. These modern techniques have been developed for situations with hundreds of predictors, and the researchers' theories do not differentiate every single variable. This situation is not very common in psychology (although examples like running PCAs on photographs of faces exist), but researchers could reconceptualize some of their problems in this way. For example, within education researchers often sum scores across a large number of items for each of several sub-tests and then use these scores in a regression. This is conceptually very similar to the PCR approach. In fact, the PCR approach would arguably be better since it would eliminate items that do not correlate well with the rest. Similarly, cognitive psychologists looking at hundreds of trials on say, memorizing word lists, could do the same. One nice thing about these approaches is that you are able to analyze data sets with which traditional methods have difficulty.

The second situation is where the high level of collinearity among several predictor variables makes using theories, which will often be about the relationships among only two or three variables, difficult. This is the more common situation within psychology, and is the situation in which researchers often use the stepwise approaches. In these cases these automated functions can help guide the researcher, but they should not dictate how model selection progresses.

The question is: are the standard stepwise approaches learned during your undergraduate courses okay for these two specific situations? For the first situation, with hundreds of predictors, the traditional approaches are not good, and the alternatives described here should be used. For the second situation, the traditional approaches *may* be okay, but each of the alternatives presented in this chapter has advantages so should be considered. The best subset regression (bsr) has an obvious advantage in that the traditional stepwise approaches are not guaranteed to reach the best, in some sense, set of predictors. The ridge and the lasso regressions have advantages over best subset regressions because, with correlated predictor variables, some coefficients may become high and unstable, and the shrinkage used in both of these procedures helps lessen this. Further, the graphs produced help show unstable coefficients. The lasso has the advantage over ridge regression in that, as well as shrinking the coefficients, some coefficients drop out, making the solution simpler. Hastie et al. (2001) describe it as 'a kind of continuous subset selection' (p. 64). As described in this chapter, PCR and PLSR seem different, but in fact principal component analysis is very closely related to multiple regression. The PCR technique has the two distinct steps: PCA and multiple regression. This division can be helpful if, for example, you wanted to use the components in some other analyses. Both of these procedures have the problem that the components are likely to include lots of variables and so not be very parsimonious if you consider all these variables, but if you can describe these summaries as meaningful indices in their own right, then it becomes simpler. This is easier to do with PCR than with PLSR because you do not need to refer to the response variable in describing the components for PCR.

An important attribute in the choice of statistical test is the ease of communicating the results to an audience. The traditional stepwise methods and best subset regressions are top of our list for communication because they have a long history in psychology, but PCR also will be easily understood by most psychologists because most of them have knowledge of PCA. Ridge and lasso regression will be more difficult to describe, but because the constraints used look simple ($\Sigma\beta_k^2 \leq k$, and $\Sigma|\beta_k| \leq k$) they should be able to be explained to most psychology audiences. There have been several extensions to the lasso, which we did not describe for reasons both of space and their complexity. PLSR is a newer technique and more difficult to describe.

Statisticians often run Monte Carlo simulation studies where they create data with known values, and see how well the techniques are at estimating these values. They have looked at the behavior of each of these methods. Another way to compare methods is with cross validation, and in fact cross validation is a method to help decide how much shrinkage to use in the ridge and lasso, how many variables to use in best subset regression, and how many components to use in PCR and PLSR.[7] Hastie et al. (2001) review these different procedures and describe how PCR, PLSR, and the ridge regression behave similarly, and they preferred ridge regression of these three. They compare the lasso and ridge, but the parsimony of the lasso makes it a clear winner in our minds. This does not mean the lasso is either going to find the optimal (in some sense) solution or that it should always be used.

---

[7]It is worth saying that different types of cross validation can be done easily with the different R functions used here. For space reasons we only briefly mention them.

For example, Zhao and Yu (2006) find it has trouble excluding non-predictive variables which are highly correlated with some predictive ones. The original lasso paper (Tibshirani, 1996) has been cited hundreds of times and spawned numerous extensions and modifications, so there are already new alternatives with more to come.

So, which procedure wins? The boring answer is that you should use several different techniques. Each makes different assumptions and has different purposes. If they all give a similar solution then there is a good chance that is a good solution. If they give different answers then you will have to think more about your specific question and the aims of your research. You may have to choose one technique.

That is the boring answer, which you probably did not want to hear. We know people do not like the advice 'try lots and see what you get', and there are some sound reasons against trying too many approaches because sometimes researchers may be biased to choose the answers which suit them best. Therefore, we will give more direct advice, but with the caveat that it usually is worth trying several techniques. On the basis of what type of output you get, the simplicity of the solution, and the ease of exposition, we recommend the lasso and PCR when you have multiple correlated predictor variables, when you lack any clear theories about the relationships among these, and you wish to see how these predictors relate to a response variable. Use the lasso when you are more interested in the response variable, and PCR when you are more interested in the predictor variables.

SOME WORDS/CONCEPTS WORTH REMEMBERING

R concepts

- printing matrices: useful for correlation matrices.

R functions

- **cbind**: combines variables to use as a group;
- **cor**: for making a correlation matrix;
- **cor.test**: for statistics correlating two variables;
- **diag**: for using the diagonal of a matrix;
- **lower.tri, upper.tri**: to use the lower and upper triangles of a matrix;
- **spm**: scatterplot matrices;
- **regsubsets**: for best subset regression;
- **which, which.max**: to identify which case is the largest in a set;
- **lm.ridge**: ridge regression;
- **lars**: least angle regression (for the lasso);
- **pcr**: principal component regression;
- **princomp**: principal component analysis;
- **plsr**: partial least squares regression.

Statistical concepts

- bsr: best subset regression;
- ridge and lasso: two methods that constrain coefficients;
- PCA: principal component analysis;
- PCR and PLS: model simplification based on PCA.

FURTHER READING

Hastie, T., Tibshirani, R. & Friedman, J. (2001). *The elements of statistical learning: Data mining, inference, and prediction.* Springer-Verlag: New York. Webpage: http://www-stat.stanford.edu/~tibs/ElemStatLearn/. Chapter 3 of this book provides a mathematical introduction to all of these procedures. Trevor Hastie says the 2nd edition should be out soon.

   The procedures within this book have been written up as an R package: Halvorsen, K. (2007) *ElemStatLearn: Data sets, functions and examples from the book: 'The Elements of Statistical Learning, Data Mining, Inference, and Prediction'* by Trevor Hastie, Robert Tibshirani and Jerome Friedman. R package version 0.1–3.

Mevlik, B-H. (2006). The *pls* package. *R News, 6/3*, 12-17. Available: http://cran.r-project.org/doc/Rnews/Rnews_2006-3.pdf. This is a good brief description of PCR and PLSR.

The lasso page, http://www-stat.stanford.edu/~tibs/lasso.html, has links to various descriptions of the lasso.

6

Generalized linear models (GLMs)

Learning outcomes

1. To be able to run generalized linear models (GLMs) for response variables that are:

    *   normally distributed;
    *   counts (frequencies);
    *   binary (like a single YES or NO question);
    *   binomial (like a proportion correct on several binary variables).

2. To be able to present the results of a GLM graphically.

One of the most important advances in statistical modeling during the last 50 years was begun by Nelder and Wedderburn (1972). They showed that linear regression could be extended to a larger set of situations, including many that are frequently encountered in psychology. Requiring the model to be linearly related to the responses and requiring normally distributed residuals are not appropriate for many research problems. In the past psychologists often still ran regressions and ANOVAs pretending that their response variables had these characteristics, but many did so with a guilty feeling. These guilty feelings lead to high levels of stress, poor health, and they died. Not really, but only because they could justify this behavior, to some extent, because alternatives were not readily available. The development of the generalized linear model (GLM) means that alternatives are now available. Different functions can be used to *link* the predicted values with a linear combination of the predictor variables. There is a notation used for GLMs that is worth introducing. The linear combination of the predictor variables (and this can include variables multiplied by each other and functions of these variables – it is linear in the β values) is called the model and is denoted with η_i. The predicted values are denoted with μ_i. The link function, denoted $g()$, connects the model and the predicted values such that $g(\mu_i) = \eta_i$.

The link functions are one of the two key concepts needed to conduct GLMs. In this chapter three link functions are considered: the identity function; the log function; and the logit function. These link functions have different error distributions usually associated with them. The error distribution is the second key concept for GLMs. The error distributions usually associated with each of the link functions are: normally distributed errors with the identity link; Poisson distributed errors with the log link; and binomially distributed errors with the logit link.

The phrase 'identity function' in mathematics means a function that maps something onto itself. For the identity function,

$$g(\mu_i) = \mu_i = \eta_i = \beta 0 + \Sigma \beta k \, xk_i$$

for k predictor variables. The μ_i are not the observed responses themselves (i.e., not the y_i), but predicted values. To get the responses we have to include an error term, so $y_i = \beta 0 + \Sigma \beta k \, xk_i + e_i$. If we assume normally distributed errors, which is the standard assumption with the identity link, this is simply the standard linear multiple regression that has been covered in previous chapters. This is $y_i = \eta_i + e_i$, where it is assumed that $e_i \sim N(0, \sigma)$, which means the e_i come from a normal distribution with a mean of 0 and an unknown standard deviation (σ) which will be estimated. This is a special case of GLM. The model part of the regression, the η_i, is linear in the sense that the βs can be separated from the Xs, and this is why this procedure is called the *generalized linear model*. As is the norm for many statistical procedures it is often referred to by its abbreviation: GLM.

A common situation in psychology is where the dependent variable is a frequency. For example, this might be how many times a child asks for help in a classroom. If these occurrences are independent from each other and are based only on a single probability for each person then often it is reasonable to assume that the data follow a Poisson distribution and the log link is appropriate: $ln(\mu_i) = \eta_i$ with error following a Poisson distribution. This is $ln(y_i) = \eta_i + e_i$, where $e_i \sim$ Poisson(λ), which means the residuals are from a Poisson distribution with an unknown mean of λ. With the Poisson distribution the standard deviation is the same as the mean. Most of the time in psychology when a Poisson distribution is used λ is small (<3), and in these cases it has high expected probabilities for low frequencies and then the expected probabilities decline as the frequencies increase (i.e., it is positively skewed). Thus, it is expected that most children ask few questions, but that some may ask lots. Figure 6.1 shows some examples of Poisson distributions for $\lambda = 1, 2, 5, 10$, and 20. As the value of λ reaches 10 and 20 the distribution looks more like a normal distribution, so in these situations people would often just assume normally distributed errors. The **lines(spline(0:40, dpois(0:40,i)))** tells R to draw lines based on a smooth curve called a **spline** (covered more in Chapter 7) with x-axis coordinates of 0–40, and y-axis coordinates corresponding to these values for $\lambda = $ **i**. The **for** function tells R to do this for $\lambda = 1, 2, 5, 10$, and 20. The locations of the '$\lambda = $' in the figure were based on trial and error with the **text** function. This would be an occasion where the **locator(1)** function could have been used. This would allow you to place the text where you wanted on the graph (try **text(locator(1), expression(lambda," = 1"),pos=4))**.

```
plot(0:40,dpois(0:40,1),xlab="Variable",ylab="",ylim=c(0,.5),
    col="white")
for (i in c(1,2,5,10,20)) lines(spline(0:40,dpois(0:40,i)))
text(1, .37, expression(lambda," = 1"),pos=4)
text(2.4, .26, expression(lambda," = 2"),pos=4)
text(4, .20, expression(lambda," = 5"), pos=4)
```

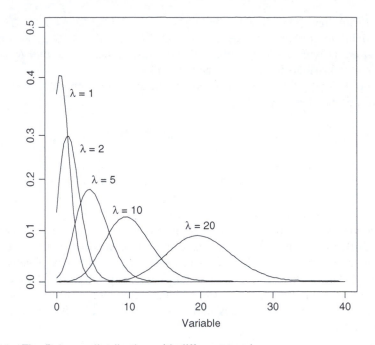

Figure 6.1   The Poisson distribution with different λ values

```
text(6.8, .14, expression(lambda,"    = 10"), pos=4)
text(17, .11, expression(lambda,"    = 20"), pos=4)
```

This procedure is sometimes called a Poisson regression. The phrase Poisson regression is usually used when you have a frequency variable for an individual person. Another situation where Poisson regressions are used is when the variable is positively skewed, which makes it look like the Poisson distribution with small values of λ, but that researchers are not claiming that the responses actually occurred by a Poisson process. Reaction time data are often modeled this way.[1] The phrase log-linear model is related and is used when you have categorical data and you are trying to model the number of people in a particular cell in a contingency table (Agresti, 2002). These models are common in sociology, but less common in psychology.

Another situation that often occurs in psychology is where a person's score is the number of correct responses out of a total (i.e., a proportion). In these situations the logit link function can be used. The logit link function is: $ln(\mu_i/[1 - \mu_i]) = \eta_i$, where ln means the natural logarithm. Logit stands for log-odds. It is usually assumed that the error is binomially distributed. This is called a logistic regression. The predicted value, μ_i, is the probability of a correct response on an item. A special case of this is when the variable is dichotomous, for example, a pupil either passes or fails a task. The error term follows a Bernoulli distribution, which is a special case of the binomial distribution when there is only a single trial. This special case deserves mention because

---

[1] In R 'Poisson' requires the data to be integer numbers, but putting 'quasipoisson' in as the family allows you to have non-integer values (with this method the dispersions are also allowed to vary, which can be a good thing, but is more complex, see Hinde & Demétrio, 1998; Wright, 1997).

logistic regressions are often used for dichotomous variables, and in fact when some people refer to logistic regression they are actually referring just to this special case. The variance of the binomial distribution is a simple function of its mean, and, as with the Poisson regression, this can be useful in some situations (Hinde & Demétrio, 1998).

Example 7 – Different types of GLMs in R

- Data: Made-up school data. Previously used in Wright (2006a).
- Research question: How does intelligence relate to several other variables.
- Purpose: To illustrate how GLMs can be used for different types of response variables and how graphs of the predicted values can be made.

We will use a very small data set to illustrate the different types of GLM before describing an example with real data. For illustrative purposes suppose that we have data on 20 children. The data (from Wright, 2006a) include values from a standardized intelligence test that is normally distributed with a mean of 0 and a standard deviation of 1. We want to see how these scores relate with scores from a scale of socializability, the number of books read, the number correct out of 10 on a math quiz, and whether the child received detention during the previous year.

```
webreg <- "http://www.sagepub.co.uk//wrightandlondon//"
glmexample <- read.table(paste(webreg,"glmexample.dat",sep=""),
   header=T)
attach(glmexample)
```

The first model we look at is a simple linear regression between social scores and intelligence scores. This regression can be done with **glm** or with **lm**, but we will use **glm** for illustration. The defaults for **glm** are to assume the residuals are normally distributed and that the link function is the identity function (i.e., none). It shows a positive and significant relationship between **test** and **social**. The output looks a little different than what you get with the **lm** function. The dispersion parameter is not usually mentioned with the standard regression because it is allowed to vary. With the other GLMs it can be more important because the standard deviation/variance is often assumed to be a function of the mean. When we wrote out the model with an error term, we wrote $e_i \sim N(0, \sigma)$. The dispersion value (1.36) is the estimate of σ^2. The residual sum of squares (24.531) divided by its degrees of freedom (18) is this value. The estimate for σ is the square root, or 1.17. The sum of squares (listed as deviance measures) and the coefficient estimates are the same as you would get with the **lm** function. Statistics like R^2 are not printed, but you can calculate this yourself: $(41.710 - 24.531)/41.710 = 41\%$.

```
socreg <- glm(social~test)
summary(socreg)

Call:
glm(formula = social ~ test)
```

```
Deviance Residuals:
     Min         1Q     Median        3Q        Max
-1.85983   -0.88971   -0.08739    1.08340    1.57728


Coefficients:
            Estimate Std. Error t value Pr(>|t|)
(Intercept)  -0.2224     0.2658  -0.836  0.41385
test          0.8743     0.2463   3.550  0.00229 **
---
Signif. codes:  0 '***' 0.001 '**' 0.01 '*' 0.05 '.' 0.1 ' ' 1


 (Dispersion parameter for gaussian family taken to be 1.362831)


    Null deviance: 41.710  on 19  degrees of freedom
Residual deviance: 24.531  on 18  degrees of freedom
AIC: 66.842


Number of Fisher Scoring iterations: 2
```

The following regression shows that there is a negative relationship for test score in predicting detention. Figure 6.2, described below, helps to show how strong the relationship is. The usual test statistic is: $t(18) = 1.90$, $p = .06$. **detent** is a binary variable meaning it is like a single coin flip. Binomial means like a number of coin flips, so binary is a special case of binomial where it is only a single coin flip. This distinction is important. Because the variance for the binomial distribution is a function of its mean the function assumes an appropriate variance for the error distribution. The **glm** procedure allows this assumption to be lifted, but this requires slightly more complex computation (see Venables & Ripley, 2002, for how to do this in R).

```
detreg <- glm(detent~test,binomial)
summary(detreg)


Call:
glm(formula = detent ~ test, family = binomial)


Deviance Residuals:
     Min        1Q     Median        3Q        Max
 -1.4593   -0.8497   -0.3505    0.9032     1.5888


Coefficients:
            Estimate Std. Error z value Pr(>|z|)
(Intercept)  -0.3385     0.5310  -0.637   0.5239
test         -1.3430     0.7059  -1.902   0.0571 .
---
Signif. codes:  0 '***' 0.001 '**' 0.01 '*' 0.05 '.' 0.1 ' ' 1


 (Dispersion parameter for binomial family taken to be 1)
```

```
    Null deviance: 26.920   on 19  degrees of freedom
Residual deviance: 21.185   on 18  degrees of freedom
AIC: 25.185

Number of Fisher Scoring iterations: 5
```

R has different ways to run regressions with proportions. Here we have entered the proportions as a two column matrix where the first column is the number of correct answers and the second column is the number of incorrect answers. This is useful in case people have answered different numbers of questions (see Venables & Ripley, 2002, for discussion). This model shows that **test** predicts **math** scores.

```
x <- cbind(math,10-math)
mathreg <- glm(x~test,binomial)
summary(mathreg)


Call:
glm(formula = x   ~ test, family = binomial)

Deviance Residuals:
    Min       1Q    Median       3Q      Max
-1.7942   -0.6700   0.2121    0.7118   1.3207

Coefficients:
              Estimate Std. Error z value Pr(>|z|)
(Intercept)   -0.3216     0.2021  -1.591    0.112
test           2.7027     0.4037   6.695 2.16e-11 ***
---
Signif. codes:  0 '***' 0.001 '**' 0.01 '*' 0.05 '.' 0.1 ' ' 1

(Dispersion parameter for binomial family taken to be 1)

    Null deviance: 138.151   on 19  degrees of freedom
Residual deviance:  14.885   on 18  degrees of freedom
AIC: 52.803

Number of Fisher Scoring iterations: 5
```

Finally, **test** also predicts the number of **books**. It is assumed the number of books follows a Poisson distribution.

```
bookreg <- glm(books~test,poisson)
summary(bookreg)


Call:
glm(formula = books ~ test, family = poisson)
```

```
Deviance Residuals:
    Min       1Q    Median      3Q       Max
-1.4813   -0.7041  -0.2727   0.2818    1.1853


Coefficients:
             Estimate Std. Error z value Pr(>|z|)
(Intercept)  -0.4047     0.3109   -1.302    0.193
test          1.1304     0.1734    6.518 7.12e-11 ***
---
Signif. codes:  0 '***' 0.001 '**' 0.01 '*' 0.05 '.' 0.1 ' ' 1


(Dispersion parameter for poisson family taken to be 1)

    Null deviance: 62.779  on 19  degrees of freedom
Residual deviance: 11.063  on 18  degrees of freedom
AIC: 47.269

Number of Fisher Scoring iterations: 5
```

The generalized linear model is important for statistics because it turns non-linear models into linear ones with the link function. This means that a computer can solve them relatively easily. The difficulty for most people is trying to conceptualize the models in terms of logits and logs. It is usually easier to see simplified patterns in data in graphical format than in numerical format, and we think this is particularly true for GLMs. Figure 6.2 shows the four graphs which correspond to the four models just discussed. The **predict(glm.object,type="response")** tells the computer to have on the y axis the predicted response values, the μ_i.

```
par(mfrow=c(2,2))
plot(test,social)
lines(test,predict(socreg,type="response"))
plot(test,detent)
lines(test,predict(detreg,type="response"))
plot(test,math/10)
lines(test,predict(mathreg,type="response"))
plot(test,books)
lines(test,predict(bookreg,type="response"))
par(mfrow=c(1,1))
```

As with other modeling procedures, we often want to compare different GLMs. Consider the example of predicting detention. Suppose we were interested in whether including the scores on **social** increased the predictive value over just using **test** on its own. The **anova** output shows that the main effect of the additional term, **social**, is non-significant ($\chi^2(1) = 1.77, p = .18$).

```
det2 <- glm(detent~test+social,binomial)
anova(detreg,det2,test="Chi")
```

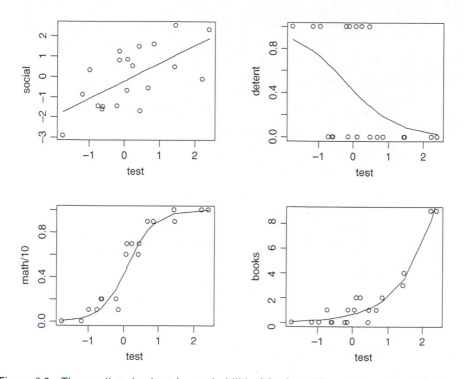

Figure 6.2   The predicted values (or probabilities) for four different generalized linear models are shown by the lines. The observed values are shown with circles

```
Analysis of Deviance Table
Model 1: detent ~ test
Model 2: detent ~ test + social

   Resid. Df Resid. Dev Df Deviance P(>|Chi|)
1         18      21.185
2         17      19.412  1    1.773      0.183
```

Next the interaction is added, and this also fails to significantly improve the fit of the model ($\chi^2(1) = 0.12$, $p = .73$). We told the computer to use **test="Chi"**. If we had said **test="F"** R would print a warning that the F test is not appropriate in this circumstance.

```
det3 <- glm(detent~test*social,binomial)
anova(det2,det3,test="Chi")

Analysis of Deviance Table

Model 1: detent ~ test + social
Model 2: detent ~ test * social
   Resid. Df Resid. Dev Df Deviance P(>|Chi|)
1         17     19.4117
2         16     19.2899  1   0.1219      0.7270
```

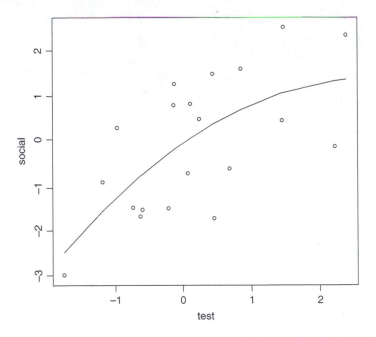

Figure 6.3   A scatterplot of social with test scores, with the predicted values of social with test scores from a quadratic regression

Finally, the code below shows that you can include things like polynomial functions within the **glm** function. Remember that the 'linear' in linear models is in terms of the β values, not the Xs. Here it includes both the linear and the quadratic (the **2** in the **poly** function) polynomial terms. It is non-significant. Figure 6.3 shows that the curve does not deviate much from the linear.

```
socialpoly <- glm(social~poly(test,2))
plot(test,social)
lines(test,predict(socialpoly,type="response"))
```

Example 8 – Finding reasonable doubt

- Data: Wright and Hall (2007, Exp. 2)
- Packages: MASS, e1071, boot
- Research question: Can we change what people think 'reasonable doubt' is?
- Purpose: To further illustrate the most common GLM (logistic regression) and to place additional stress on preliminary data analysis.

This example comes from Wright and Hall (2007) who were interested in a reasonable doubt instruction on juror decision making. Participants read a brief crime summary and were asked to render a verdict. They were told to tick 'guilty' if they believed the defendant's guilt was 'beyond a reasonable doubt.' They also rated their belief in guilt on a 0.00 to 1.00 probability scale. Participants were either in a control condition where they received no further instructions or they were in an experimental condition where they received a more elaborate instruction which had two predicted effects. First, it was expected to lower belief in guilt and second it was expected to lower the reasonable doubt threshold. 172 participants made both a belief in guilt judgment and a binary verdict (guilty vs. not guilty). It seems plausible to define 'reasonable doubt' as where there was a 50% chance on the probability scale for rendering a guilty verdict. This is called LD50, for lethal dose 50%, and comes from medicine where analysts have used it to estimate the dose of a drug where 50% of the animals would be expected to die. Most medical testing no longer uses this procedure, but the name has stuck (and serves as a reminder of how ethical principles for science have evolved). From the estimates of a logistic regression LD50 is $-\beta0/\beta1$. If run on the two groups separately the confidence interval can be found with the **dose.p** function from the **MASS** library. This is what Wright and Hall used in their paper, but the confidence intervals could also be found with bootstrapping.

One purpose of this example is to go through in more detail how a researcher would analyze data. Because there are only three variables, this is much simpler than most datasets. The first command, **read.table**, accesses the data and then you must **attach** them. The **names** command shows that the variable names are all in capitals. **GUILTY** and **FORM** are binary variables. For **GUILTY**, 0 is not guilty and 1 is guilty, and **FORM** is either 0 for the control group or 1 for the imagine instruction group.

```
webreg <- "http://www.sagepub.co.uk//wrightandlondon//"
juryrd <- read.table(paste(webreg,"juryrd.dat",sep=""),
   header=T)
attach(juryrd)
names(juryrd)
```

```
[1] "SUBNO" "GUILTY" "FORM" "BELIEF"
```

The variable **BELIEF** is important because we want to see if it is affected by the condition, and also we want to see how it relates to verdict. The left panel of Figure 6.4 shows the default histogram of this variable. It is negatively skewed.

```
par(mfrow=c(1,2))
hist(BELIEF)
```

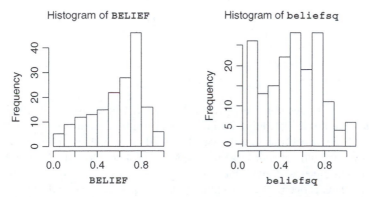

Figure 6.4   The left panel shows a histogram of the untransformed variable **BELIEF**, showing a negative skew. The right panel shows a histogram of the variable transformed with **BELIEF^2**

To check skewness, we used the **skewness** function in the library **e1071**. If you did not know where a skewness function was you would type **help.search("skewness")** and it would find a couple of functions for you. A rough estimate of the error of skewness is $\sqrt{6/n}$ so an estimate of the 95% confidence interval is presented below this.

```
library(e1071)
skewness(BELIEF)


[1] -0.7808726


lb <- format(skewness(BELIEF) - sqrt(6/length(BELIEF))*1.96,digit=3)
ub <- format(skewness(BELIEF) + sqrt(6/length(BELIEF))*1.96,digit=3)
print(paste("Lower Bound = ", lb, "Upper Bound = ", ub))


[1] "Lower Bound = -1.15 Upper Bound = -0.415"
```

As discussed in Chapter 2, an alternative is to calculate bootstrap estimates of this. These are fairly similar to the one created above. The BCa intervals are usually preferred and here the 95% BCa CI goes from -1.07 to -0.53.

```
library(boot)
beliefboot <- boot(BELIEF,function(x,i) skewness(x[i]),R=1000)
boot.ci(beliefboot)


BOOTSTRAP CONFIDENCE INTERVAL CALCULATIONS
Based on 1000 bootstrap replicates


CALL :
boot.ci(boot.out = beliefboot)


Intervals :
Level       Normal               Basic
95%    (-1.0507, -0.5155 )   (-1.0504, -0.5083 )


Level       Percentile             BCa
95%    (-1.0534, -0.5114 )   (-1.0653, -0.5336 )
Calculations and Intervals on Original Scale
Warning message:
bootstrap variances needed for studentized intervals in:
boot.ci(beliefboot)
```

The 95% BCa CI does not overlap with 0 so we can be confident that the variable's distribution would be skewed if we had tested an infinite number of people from the population from which this sample was drawn (they were University of Bristol [UK] psychology students). Of course 'an infinite number of people from the population from which this sample was drawn' does not exist (there were about 300 of them in any given year). This is one of the conceptual problems people have to deal with when first learning about hypothesis testing and confidence intervals (Wright & London, 2009), but it is worth providing occasional reminders about the difficulties in making inference with all statistics.

Because an assumption of many statistical tests is that the residuals are normally distributed, not just symmetrical, it is often worth testing this. The two most used tests are the Shapiro-Wilks test and the Kolmogorov-Smirnov test. Here, both of these are statistically significant. The Shariro-Wilks test is usually preferred. The code and output for both of these tests are:

`shapiro.test(BELIEF)`

```
Shapiro-Wilk normality test

data: BELIEF
W = 0.9271, p-value = 1.291e-07
```

`ks.test(BELIEF,"pnorm")`

```
One-sample Kolmogorov-Smirnov test

data: BELIEF
D = 0.5502, p-value < 2.2e-16
alternative hypothesis: two-sided
Warning message:
cannot compute correct p-values with ties in: ks.test
    (BELIEF, "pnorm")
```

When data are negatively skewed people often try squaring the variable, and as can be seen below this works fairly well. The right panel of Figure 6.4 shows **hist(beliefsq)**. This transformed variable is fairly symmetric. As can be seen, the BCa 95% CI overlaps with zero.

```
beliefsq <- BELIEF^2
hist(beliefsq)
par(mfrow=c(1,1))
bboot2 <- boot(beliefsq,function(x,i) skewness(x[i]),R=1000)
boot.ci(bboot2)


BOOTSTRAP CONFIDENCE INTERVAL CALCULATIONS
Based on 1000 bootstrap replicates

CALL :
boot.ci(boot.out = bboot2)

Intervals :
Level       Normal                 Basic
95%    (-0.2971,  0.1458 )    (-0.2950,  0.1516 )

Level       Percentile             BCa
95%    (-0.3170,  0.1296 )    (-0.3130,  0.1359 )
Calculations and Intervals on Original Scale
```

```
Warning message:
bootstrap variances needed for studentized intervals in:
boot.ci(bboot2)
```

Below is the *t* test in regression format, which assumes equal variances, so is equivalent to the *t* test in SPSS. In R the function **t.test**, however, does not make this assumption. Here, the standard deviations are .25 and .23, so are incredibly close (found by typing **sd(wbeliefsq[[1]])** and **sd(wbeliefsq[[2]])** below), so these methods give almost exactly the same values. If we reject the H0, as most people would since $p < .05$, it means we accept that the instruction does affect people's beliefs in guilt.

```
reg1 <- lm(beliefsq~FORM)
summary(reg1)

Call:
lm(formula = beliefsq ~ FORM)

Residuals:
     Min      1Q   Median       3Q      Max
-0.46176 -0.21176  0.02824  0.17824  0.51573

Coefficients:
            Estimate Std. Error t value Pr(>|t|)
(Intercept)  0.46176    0.02572  17.952  <2e-16 ***
FORM        -0.07500    0.03638  -2.062  0.0408 *
---
Signif. codes:  0 '***' 0.001 '**' 0.01 '*' 0.05 '.' 0.1 ' ' 1

Residual standard error: 0.2385 on 170 degrees of freedom
Multiple R-Squared: 0.02439,    Adjusted R-squared: 0.01865
F-statistic:  4.25 on 1 and 170 DF,  p-value: 0.04076
```

The following code shows that the default **t.test** in R produces something very slightly different. If you want **t.test** to assume equal variances include **var.equal=T**.

```
t.test(beliefsq~FORM)

Welch Two Sample t-test

data:  beliefsq by FORM
t = 2.0617, df = 168.244, p-value = 0.04078
alternative hypothesis: true difference in means is not equal to 0
95 percent confidence interval:
 0.003183091 0.146812839
sample estimates:
mean in group control mean in group imagine
            0.4617631             0.3867651
```

The **split** command below creates a variable for the belief squared variable that is separated by the groups of **FORM**. This is needed for some procedures, like the Wilcoxon, and useful for others. The Wilcoxon is a test from the 1950s that still is often used for ranked data. Here it compares the ranks for the belief variables for the two groups. Wow, look at that p value ... because both authors teach statistics we often try to find examples like this which are incredibly close to that magical .05. We are far more pleased than is healthy! It is also interesting because there are different ways to calculate the Wilcoxon rank sum test (which is equivalent to the Mann-Whitney U). Here, the continuity correction is used. If it is not used, the p value becomes a tiny bit smaller ($p = .04981$). However, R calculates these differently than the original Wilcoxon paper, than the way it is done in many introductory textbooks (like Wright & London, 2009), and than even S-Plus. If the other method is used, then $p = .0507$. Of course the difference between .049 and .051 should be of no interest to anyone, but sadly it is.

```
wbeliefsq <- split(beliefsq, FORM)
wilcox.test(wbeliefsq[[1]],wbeliefsq[[2]])

Wilcoxon rank sum test with continuity correction

data: wbeliefsq[[1]] and wbeliefsq[[2]]
W = 4336, p-value = 0.04999
alternative hypothesis: true mu is not equal to 0
```

Now we can do the logistic regressions. Much of the time conducting statistics will be preparing the data and preliminary exploratory analyses. We begin with the belief squared variable to predict the verdict. Next we add the experimental condition. The **anova** command compares these two models. It gives us a χ^2 value which we can either look up in a table, or do as we have done here and let R calculate the associated p value. We could have written: **pchisq(15.351,1,lower.tail=F)**. Either way, this shows it is statistically significant, meaning that controlling for **beliefsq**, **FORM** has an effect ($8.9e-05$ means 0.000089). If we had written **anova(reg2,reg3,test="Chi")**, it would have printed the appropriate p value, too.

```
reg2 <- glm(GUILTY~beliefsq, binomial)
reg3 <- glm(GUILTY~beliefsq+FORM, binomial)
anova(reg2,reg3)

Analysis of Deviance Table

Model 1: GUILTY ~ beliefsq
Model 2: GUILTY ~ beliefsq + FORM
  Resid. Df Resid. Dev  Df Deviance
1       170    153.658
2       169    138.306   1   15.351


1-pchisq(15.351,1)

[1] 8.927366e-05
```

We then look at the values for this regression, although they are easier to see in graphical form, which is done below. This does show that there is a large (and predictable) effect for **beliefsq**, meaning that if you believe the person is guilty you are more likely to deliver a guilty verdict. There is also an effect for **FORM**. Those in the experimental group have a higher probability of making a guilty verdict once **beliefsq** has been controlled for.

```
summary(reg3)

Call:
glm(formula = GUILTY ~ beliefsq + FORM, family = binomial)

Deviance Residuals:
    Min       1Q    Median       3Q       Max
-2.5267   -0.6269   -0.1827    0.6119    2.1203

Coefficients:
            Estimate Std. Error z value Pr(>|z|)
(Intercept)  -5.6442     0.8822  -6.398 1.58e-10 ***
beliefsq      9.7443     1.4747   6.608 3.90e-11 ***
FORM          1.7436     0.4836   3.605 0.000312 ***
---
Signif. codes:  0 '***' 0.001 '**' 0.01 '*' 0.05 '.' 0.1 ' ' 1

(Dispersion parameter for binomial family taken to be 1)

    Null deviance: 235.08  on 171  degrees of freedom
Residual deviance: 138.31  on 169  degrees of freedom
AIC: 144.31

Number of Fisher Scoring iterations: 5
```

Next, the model is plotted. Because there are two groups, for drawing the separate lines it was necessary to split the **BELIEF** and the prediction variables, and draw these as two separate lines. The original **BELIEF** variable is used rather than the transformed one because this will be in an easier scale for people to understand. The **lines** function requires the variables to be ordered. This is done with the **order** function in the code below. This records the order of the belief variable for each group and then the **lines** function says to plot the curve in this order. A dashed horizontal line at 50% probability of giving a guilty verdict (the LD50 line) and vertical lines where this LD50 line touches the curves for the two groups have been added using the **abline** function. These values (which are also printed with the **paste** function) are found with: **sqrt(-(reg3$coef[1]/reg3$coef[2]))**. Figure 6.5 is the resulting graph. You would usually annotate the graph with words using the **text**, **paste**, and/or **expression** functions.

```
plot(BELIEF,predict(reg3,type="response"),ylab="Probability of a
    guilty verdict")
beliefs <- split(BELIEF,FORM)
```

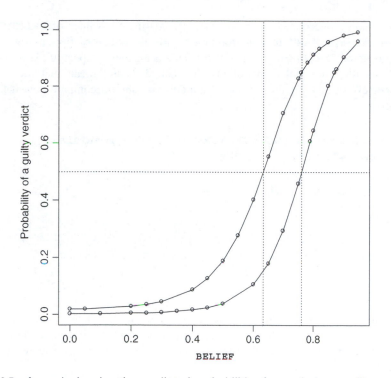

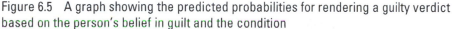

Figure 6.5   A graph showing the predicted probabilities for rendering a guilty verdict based on the person's belief in guilt and the condition

```
predicts <- split(predict(reg3,type="response"),FORM)
o1 <- order(beliefs[[1]])
o2 <- order(beliefs[[2]])
lines(beliefs[[1]][o1],predicts[[1]][o1])
lines(beliefs[[2]][o2],predicts[[2]][o2])
abline(h=.5,lty=3)
ld1 <- sqrt(-(reg3$coef[1]/reg3$coef[2]))
ld2 <- sqrt(-((reg3$coef[1]+reg3$coef[3])/reg3$coef[2]))
abline(v=ld1,lty=3)
abline(v=ld2,lty=3)
paste("Control LD50 =", format(ld1,dig=2), "Instruction LD50 =",
    format(ld2,dig=2))
```

```
[1] "Control LD50 = 0.76 Instruction LD50 = 0.63"
```

For completeness it is worth checking the interaction between **beliefsq** and **FORM**. The interaction is significant which has several possible interpretations (many models with one additional *df* fit better than the previous model, the interaction is just one of them … more on this in Chapter 7). The slope is steeper for the instruction group so one interpretation is that the variance in the interpretation of reasonable doubt is less for this group. This is presumably a good thing since you would want potential jurors to have the same meaning of 'reasonable doubt' as each other.

It could also be interpreted as the effect only existing for values of **BELIEF** above about .4. As several different interpretations are all valid descriptions of the data, caution is urged accepting any one of them. This is a general aspect of many statistical procedures; showing that a model fits the data does not mean that the model is correct or even good. To show a model is good you should compare how it fits with alternatives.

```
anova(reg3,glm(GUILTY~FORM*beliefsq,binomial),
   test="Chi")

Analysis of Deviance Table

Model 1: GUILTY ~ beliefsq + FORM
Model 2: GUILTY ~ FORM * beliefsq
  Resid. Df Resid. Dev  Df Deviance P(>|Chi|)
1       169     138.306
2       168     133.177   1    5.130      0.024
```

We will do one final bit of analysis: the standard chi-square test, here between **GUILTY** and **FORM**. First we look at the contingency table and then we run the **chisq.test**. The **correct=F** means Yates' correction factor is not used. If you use the correction factor you get $\chi^2(1) = 1.92, p = .17$.

```
table(GUILTY,FORM)

FORM
GUILTY  0  1
     0 54 44
     1 32 42
```

```
chisq.test(GUILTY,FORM,correct=F)

Pearson's Chi-squared test

data: GUILTY and FORM
X-squared = 2.3718, df = 1, p-value = 0.1235
```

This can also be run as a **glm** (as a log-linear model) but this requires creating variables for the cells of the table above:

```
cells <- c(54,44,32,42)
imagine <- c(0,1,0,1)
guilty <- c(0,0,1,1)
glm(cells~imagine+guilty,poisson)

Call: glm(formula = cells ~ imagine + guilty, family=
   poisson)
```

```
Coefficients:
(Intercept)        imagine           guilty
    3.892e+00      5.414e-16      -2.809e-01

Degrees of Freedom: 3 Total (i.e. Null);  1 Residual
Null Deviance:          5.737
Residual Deviance:  2.378            AIC: 30.72
```

and you get the likelihood ratio Chi-square value, `2.378`. The two values are different because the two statistics are calculated differently: Pearson's Chi-square, $\Sigma(O_{ij} - E_{ij})^2/E_{ij}$ and likelihood ratio Chi-square, $2 \, \Sigma O_{ij} \, ln \, (O_{ij}/E_{ij})$.

MEDIATOR WITH GLMs

Most of the extensions described in this book can be combined. For example, if you have several predictor variables and a binary response variable, you might have the same model selection issues described in Chapter 5, but with a GLM. This is possible, and Park and Hastie (2007a, b) have developed an R package, **glmpath**, to do this. Similarly, this juror decision making example seems an interesting candidate for mediator analysis because of having one experimental variable and two dependent variables. One question could be: does the influence of the instruction affect verdict only through belief? Finding an interaction does not answer this. In the language of mediation, an interaction effect is often called a *moderator*. We say 'interesting', because the experimental manipulation does not have a significant bivariate effect on verdict, and according to some this means you should not explore mediation. But after accounting for belief, the manipulation does have an effect. This means the resulting graph looks a lot different than many mediation models (we discuss this in the caption for Figure 6.6).

The function written for mediation for linear regression in Chapter 4 can be slightly modified here (just replacing **lm** with **glm** for **reg0** and **reg2**, and assigning the correct link function – and we changed the function name to **mediatorbin** just so we would not get confused, and ran it). The command:

```
mediatorbin(FORM,GUILTY,beliefsq)
```

produces Figure 6.6. The **mediator** function in Chapter 4 can be easily altered so that it can be run for lots of different situations. Preacher and Hayes (2004) have been producing similar functions with some software packages, so we expect someone will do this for R in the near future.

SUMMARY

John Hoffman (2004: viii) writes: 'We are most fortunate to be living in a time when …'. It sounds like one of those university entrance essays where they ask you to complete the phrase and 99% of candidates provide some bland tear-jerk story about how humanity can save itself from all the evils of Gomorrah. John is not like the 99%: 'We are most fortunate to be living in a time when the statistical tools for analyzing regression models

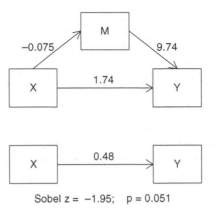

Figure 6.6 The graph for a mediation analysis for Wright and Hall (2007). The graph shows that the instruction (X) lowers the belief in guilt (M) with a coefficient of −0.075 (this is the squared variable units). The mediator has a positive influence on verdict (Y) with a coefficient of 9.74. Once the belief has been included in the model, the X is a significant positive predictor. Because the coefficients are from different types of regression, a lot of care is needed interpreting them. And the Sobel test has a p value of .051

no longer require that dependent variables follow a continuous, normal distribution.' Yes! Another member of our 'get excited by $p = .04999$ club'? While Hoffman stops short of saying using GLMs is an alternative to Buddhism's eight-fold path to nirvana, he is saying that they are useful for many statistical analyses where previously psychologists had been wrongly applying linear models on proportions, counts, and binary data. With supervisors, reviewers, and editors complaining, the timid analyst would cower beneath the excuse that there were not convenient procedures to model these. Now researchers can proudly stand up and run their generalized linear models!

SOME WORDS/CONCEPTS WORTH REMEMBERING

R concepts

- plotting predicted values from GLMs.

R functions

- `glm`: for generalized linear models;
- `shapiro.test`: to conduct Shapiro-Wilk test;
- `ks.test`: to conduct Kolmogorov-Smirnov test;
- `wilcox.test`: to conduct Wilcoxon tests;

- **pchisq**: to find p value associated with χ^2 distribution;
- **table**: to produce contingency tables;
- **chisq.test**: to conduct χ^2 test.

Statistical concepts

- GLM: generalized linear model;
- link: how the response is transformed;
- logit: the log odds;
- logistic regression: regression with proportions;
- Bernoulli: the distribution for single coin flips;
- Binomial: the distribution for multiple Bernoulli trials;
- Poisson: a distribution usually assumed for count data.

FURTHER READING

Hoffmann, J. P. (2004) *Generalized linear models: An applied approach*. Boston, MA: Pearson Education Inc. This is a good introduction and is at about the right level for graduate psychology students. We hope he reads (and likes) the final paragraph of this chapter!

Books with a bit more statistical orientation

Agresti, A. (2002) *Categorical data analysis* (2nd Edition). Hoboken, NJ: John Wiley & Sons. The bible of categorical data analysis. The examples are mostly from the social sciences. Some parts can be a bit mathematical, but you can read around these sections if necessary.

Collett, D. (2003) *Modelling binary data* (2nd Edition). Boca Raton, FL: Chapman & Hall/CRC. This book focuses on logistic regression (for both binary and binomial data). It includes some statistical bits, but a lot of it will be readable by non-statisticians. It is very good on diagnostics. A lot of the examples are from the biological sciences.

Faraway, J. J. (2006) *Extending the linear model with R: Generalized linear, mixed effects and nonparametric regression models*. Boca Raton, FL: Chapman & Hall/CRC. 'Extending the linear model' is a great title and might have been on the cover of our book if it were not for Julian Faraway beating us to it. This is a good book which covers similar material to our book, but at a more advanced level.

McCullagh, P. & Nelder, J. A. (1989) *Generalized linear models* (2nd edition). London: Chapman and Hall. This is the definitive GLM book, by some of the people most responsible for its development. It is mathematically more complex than the others in this list.

7

Regression splines and generalized additive models (GAMs)

Learning outcomes

1. To understand why you would want to use splines.
2. To know how to include different regression splines;

   - different degree polynomials; and
   - different numbers of knots.

3. To understand the basic concepts of the generalized additive model.

Estimating the linear model, $Y_i = \beta 0 + \beta 1 X 1_i + \ldots + \beta k X k_i + e_i$, is at the core of many of the statistics conducted today and is the basis for all the procedures in this book. If you allow the individual X variables to be products of themselves and other variables, then the linear model is appropriate for factorial ANOVAs and polynomial regressions, as well as estimating the mean, t-tests, etc. The flexibility of the linear model has led the authors of some textbooks and software to call this the *general linear model*. We try to avoid this phrase because it can be confused with the *generalized linear model* (GLM) described in Chapter 6.

The focus of this chapter is on extending the linear model into an additive model (see Hastie & Tibshirani, 1990; Wood, 2006 for details). In the linear model each X variable is multiplied by a scalar, the β value. This is what makes it a linear model, but this restricts the relationship between any X variable and Y (conditioned on all the other Xs). With additive models the β values are replaced by usually fairly simple (in terms of degrees of freedom) functions. The model can be re-written as: $Y_i = \alpha + f1(X1_i) + \ldots + fk(Xk_i) + e_i$. The functions are usually assumed to be *splines* with a small number of *knots*. More complex functions can be used, but this may cause the model to overfit the observed data and thus not generalize well to new data sets. The typical graphical output shows the functions and the numeric output shows the fit of the linear and nonlinear components. The choice of functions, which often comes down to the type and complexity of the splines, is critical.

WHAT ARE SPLINES?

There are several different definitions for 'splines.' Originally, 'splines' referred to the strips of material that draftsmen used to draw curves. They would fix the spline at certain points, which they called *knots*, and the strip of material would form a smooth curve between these knots. With computers, these physical splines are less often used, but with computers mathematical splines are common. In these cases splines are often used to connect many knots with a smooth curve that goes through all the points (this is how **spline** works with the **lines** function, which is why the order of the points is important). Statisticians use what are sometimes called *smoothing splines* or *regression splines* which do not connect all the points, but instead fit a fairly simple (in terms of degrees of freedom) smooth function through the data. In this chapter regression splines will be used. These are pieces of polynomial curves connected together so that they appear smooth.

There are different types of splines used in statistics. While all are mathematically complex, we will use one that is simpler than most, is flexible, and fairly easy to understand in R. It is called a B-spline and is found using the **bs** function in the library **splines**, which is installed as part of the main R program. The functions **gam** (Hastie, 2008) and **mgcv** (Wood, 2006) are also often used and these are necessary for more complex models, but not those usually used in psychology.

The purpose of regression splines is to draw a curve through a scatterplot. One type of curve that is often used in regressions is a polynomial. Polynomials have different degrees. A 0 degree polynomial is a constant, $y_i = \beta 0$. A 1 degree polynomial is a straight line and corresponds to the simple linear regression, $y_i = \beta 0 + \beta 1 x_i$. A 2 degree polynomial is a quadratic regression of the form, $y_i = \beta 0 + \beta 1 x_i + \beta 2 x_i^2$. A 3 degree polynomial is a cubic regression, $y_i = \beta 0 + \beta 1 x_i + \beta 2 x_i^2 + \beta 3 x_i^3$. The intercept, $\beta 0$, is found with all these regressions. The number of additional β values estimated is a measure of the complexity of the model. It is called the number of degrees of freedom for that variable. So, a 2 degree polynomial has 2 degrees of freedom. Figure 7.1 shows some data and regressions for polynomials of 0, 1, 2 and 3 degrees.

```
set.seed=2007
x <- (1:100)/50
y <- 4 + 2*x - 3*x^2 + x^3 + rnorm(100,0,.3)
par(mfrow=c(1,4))
plot(x,y, main="0 degree", xlab="", ylab="", pch=19, axes=F)
box()
abline(h=mean(y),lwd=1.5)
for (i in 1:3) {
plot(x,y, main=paste(i,"degree"), xlab="", ylab="", pch=19, axes=F)
```

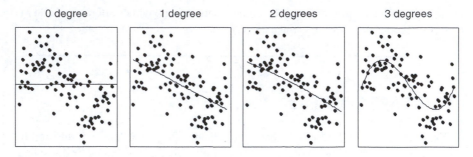

Figure 7.1   Polynomial regressions of 0, 1, 2 and 3 degrees. These correspond to a constant, a linear regression, a quadratic regression, and a cubic regression

```
box()
lines(x,predict(lm(y~poly(x,i))),lwd=1.5)}
par(mfrow=c(1,1))
```

Two aspects of the code need to be explained. The first is that a new function, **poly**, is used. This sets up polynomial contrast according to the degree that you put in. If you want a quadratic regression you write **lm(y~poly(x,degree=2))** or in shorthand **lm(y~poly(x,2))**. The second aspect is due to **poly** not allowing **degree=0**. Because of this the first part of the code makes the 0 degree graph and the remainder are made by 3 loops of the **for** command.

The data in Figure 7.1 were made using a cubic equation plus random error so we would expect the cubic regression in the right panel of Figure 7.1 to perform best. The 0 degree polynomial (the mean) clearly does not capture the data. The 1 degree polynomial (straight line) fails to capture many of the high and low points and the 2 degree polynomial (quadratic) looks similar to the 1 degree polynomial so also fails to capture these points. The 3 degree (cubic) does appear to capture the data the best. The **anova** function can also be used with these models, and the results confirm what is visible in Figure 7.1.

```
anova(lm(y~1),lm(y~poly(x,1)),lm(y~poly(x,2)),lm(y~poly(x,3)))

Analysis of Variance Table

Model 1: y ~ 1
Model 2: y ~ poly(x, 1)
Model 3: y ~ poly(x, 2)
Model 4: y ~ poly(x, 3)
  Res.Df      RSS Df Sum of Sq        F     Pr(>F)
1     99 18.5660
2     98 12.2077  1    6.3583 68.4588 7.452e-13 ***
3     97 12.1816  1    0.0260  0.2804   0.5977
4     96  8.9163  1    3.2654 35.1576 4.777e-08 ***
---
Signif. codes:  0 '***' 0.001 '**' 0.01 '*' 0.05 '.' 0.1 ' ' 1
```

The problem with polynomial regressions is that they are fairly inflexible and seldom fit the entire data set well. You can increase the complexity of the polynomial by increasing its degree, but this has two associated problems. First, there are few psychological relationships which you would predict would have this complex a function. Second, polynomials tend to shoot up or down at the ends and to be very influenced by points at the ends. The solution to this is to add together pieces of these polynomials. This is the basis of splines.

The default for the splines that we will use estimates 1 curve for the first half of the data and one for the second half. This could be done by running separate regressions. For example, for the data in Figure 7.1, we could run a linear regression for the first half:

```
qreg1 <- lm(y[1:50]~poly(x[1:50],1))
```

and for the second half:

```
qreg2 <- lm(y[51:100]~poly(x[51:100],1))
```

The plot of these is shown in the left panel of Figure 7.2.

```
par(mfrow=c(1,2))
plot(x,y, main="Piecewise linear", xlab="", ylab="", pch=19, axes=F)
box()
lines(x[1:50],predict(qreg1),lwd=1.5)
lines(x[51:100],predict(qreg2),lwd=1.5)
```

The problem is these lines do not touch. This suggests that there is some sudden shift in the relationship between the variables at this point. As shifts like this are unlikely (which is one reason why we argued against median splits in Chapter 3), it would be an advantage to make the curves meet. This is what splines do, and is where the mathematics become complex. The splines not only make the curves meet at a single point, called a *knot*, but they make them meet as smoothly as possible. If the two curves are quadratic then they form a straight line at the knot. If the curves are cubic then they form a quadratic at the knot. In R there is a package called **splines** which contains the **bs** function, which we will use to generate splines in this chapter. There are other spline functions, but

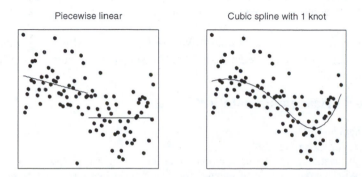

Figure 7.2   The left panel shows separate linear regressions for the first and second halves of the data. The right panel shows a cubic spline with a single knot at the median (the default location for a single knot)

the **bs** function will suffice for most purposes. Like the **poly** function, you need to tell **bs** what **degree** polynomial you have. You also need to tell it either how complex the spline is with the **df** = option, or the location of the knots with **knots = c(knot1, knot2, ...)**. A very clever aspect of splines is that each additional curve requires only one additional **df**, so that increasing the number of knots does not greatly increase the complexity of the model in terms of degrees of freedom. This is because the curves are constrained to fit together smoothing. Thus, if we want two cubic (**degree=3**) curves with a single knot it requires only **df=4**. The right panel of Figure 7.2 shows this:

```
plot(x,y, main="Cubic spline with 1 knot", xlab="", ylab="",
    pch=19, axes=F)
box()
library(splines)
lines(x,predict(lm(y~bs(x,degree=3,df=4))),lwd=1.5)
par(mfrow=c(1,1))
```

The default for **bs** is **degree=3** (cubic) without any knots. If you add one knot by making **df** equal to **degree** +1 then its default is to place the knot at the median of the variable. If you have two knots then the knots are placed at the first and second tertiles, etc. If you want to place the knots at different values this can be done with the **knots** option. This and other aspects of the **bs** function are illustrated in the next example with a small data set. After this two examples with larger data sets illustrate the typical use of these splines.

The default smoothing methods are heavily dependent on the data rather than a theory, which means that with large amounts of high quality data they can be useful exploratory tools, but with smaller samples (as typical in much psychology research) they need to be used cautiously. We believe their most valuable uses in psychology are as follows.

1. When you do not want to make any assumptions about the relationship between a covariate and the response variable. With the typical ANCOVA, it is assumed that the covariate is a linear predictor. It may be appropriate to relax this assumption using a spline of the covariate. We call this a GAMCOVA (Wright, 2008).
2. To test if a relationship is not linear.
3. As an exploratory graphical tool to plot curves within a scatterplot to search for patterns, but be careful, particularly at the ends of the scales, not to over-interpret sudden changes.

We examine these uses in our examples. We are less concerned with the exact form of the spline than is typically the case in either the statistics literature or the fields with larger amounts of data.

Example 9 – Academics' salaries

- Data: Thornton (2007).
- Package: **splines**.
- Research question: In real terms, are your professors' salaries increasing?
- Purpose: To illustrate some simple generalized linear models with a small data set.

We will begin with an example to illustrate the concepts. The data come from Thornton (2007: 22) on university faculty salaries in the US over the past three decades. The data

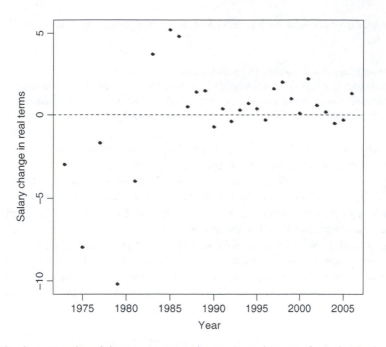

Figure 7.3   A scatterplot of the percentage change in real terms of academics' salaries by year. The dashed line is for 0%, or no change

show the amount of increase/decrease in salary in real terms (i.e., taking into account inflation). The years are:

```
years <- c(1973,1975,1977,1979,1981,1983,1985,1986,
    1987,1988,1989,1990,1991,1992,1993,1994,1995,1996,
    1997,1998,1999,2000,2001,2002,2003,2004,2005,2006)
```

and the change percent in mean salary was:

```
salary <- c(-3.0,-8.0,-1.7,-10.2,-4.0,3.7,5.2,
    4.8,0.5,1.4,1.5,-0.7,0.4,-0.4,0.3,0.7,0.4,-0.3,1.6,
    2.0,1.0,0.1,2.2,0.6,0.2,-0.5,-0.3,1.3)
```

The first step is to create a scatterplot and then to add the curves corresponding to different models to it. Figure 7.3 shows data.

```
plot(years,salary,xlab="Year",ylab="Salary change in
    real terms",pch=20)
abline(h=0,lty=2)
```

From Figure 7.3 it is clear that in the late 1970s academic salaries did not keep pace with the high rate of inflation. Usually time series methods (Chatfield, 2003; Shumway & Stoffer, 2006) are used to model data like these, particularly if you have lots of data for each year. Here, regression splines will work well to compare potential models that may account for the pattern in these data. The models that will be used are as follows.

1. Linear: That a single straight line accounts for the data.
2. Linear with one knot: That two straight lines are needed to account for the data.
3. Quadratic: That a simple curve (U or inverse U shape) accounts for the data.
4. Quadratic with one knot: That two simple curves account for the data.

The first model, **salary1**, is just the simple linear regression between the two variables, the same as **lm(salary~years)**. To make the parallel with the remaining examples clearer, we will write this as **lm(salary~bs(years, degree=1))**. The coefficients from these two models are not the same as if we had run **lm(salary~years)**, but this is because of the way **bs** works. For this simple model **bs(years,degree=1)** takes the **years** variable and re-scales it so the values go between 0 and 1. With more complicated splines it gets more difficult to match the coefficient estimates onto something meaningful with respect to the original variables. The **bs** function scales the coefficients to between 0 and 1 and makes them orthogonal to each other. Therefore, the norm is to look at these graphically and to compare models with **anova**.

```
salary1 <- lm(salary ~ bs(years,degree=1))
summary(salary1)

Call:
lm(formula = salary ~ bs(years, degree = 1))

Residuals:
    Min      1Q  Median      3Q     Max
-8.2945 -0.9241 -0.0241  0.9511  6.2289

Coefficients:
                       Estimate Std. Error t value Pr(>|t|)
(Intercept)              -2.782      1.268  -2.194   0.0374 *
bs(years, degree = 1)     4.821      2.001   2.409   0.0234 *
---
Signif. codes:  0 '***' 0.001 '**' 0.01 '*' 0.05 '.' 0.1 ' ' 1

Residual standard error: 2.97 on 26 degrees of freedom
Multiple R-Squared: 0.1825,      Adjusted R-squared: 0.151
F-statistic: 5.804 on 1 and 26 DF,  p-value: 0.02337
```

The first panel of Figure 7.4 shows this linear model. **fitted(salary1)** means the predicted values from this model. **predict(salary1)** and **salary1$fitted.values** will produce the same values.

```
par(mfrow=c(2,3))
plot(years,salary,xlab="Years",ylab="Salary change",pch = 19)
abline(h=0,lty=2)
lines(years, fitted(salary1))
```

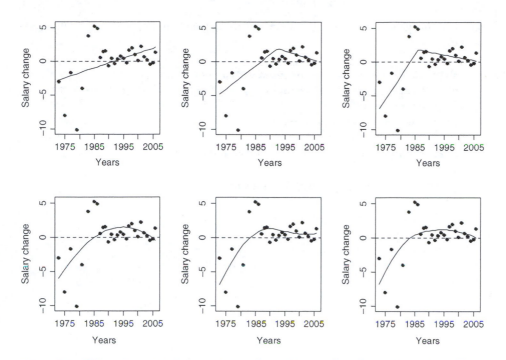

Figure 7.4   Different models for the change in real terms of academics' salaries with years (Thornton, 2007). The top panel shows a single linear model, a piecewise linear model with a knot at the median (1992.5 years), and a piecewise linear model with a knot at 1986. The bottom panel shows quadratic models with: no knots, a knot at the median, and a knot at 1986

Clearly the single straight line fails to account for the data. It overestimates change in salary during the 1970s and it is unlikely that the percentage increase in real terms will continue rising at this rate in the coming decades. The next step is to slightly increase the complexity of the model. We will do this by assuming the model is fit by two straight lines joined at a single point. This single point is called a knot. The default within the **bs** function is to place the knot at the median.

```
salary2 <- lm(salary ~ bs(years,degree=1,df=2))
plot(years,salary,xlab="Years",ylab="Salary change",pch = 19)
abline(h=0,lty=2)
lines(years,fitted(salary2))
```

If you have two knots it will set them at the 33rd and 67th percentiles. If you have a specific x value where you want to place the knot you can use the **knots** option. For example, if you thought *a priori* there might be a change at 1986 because that was when some changes in contract legislation occurred, you could write: **lm(salary~bs(years,degree=1,knots=1986))**. If you want two knots then you could write, for example, **lm(salary~bs(years,degree=1, knots=c(1980,1986)))**. The model which assumes that the trajectory of academics' salaries changed in 1986 is plotted in the third panel of Figure 7.4 and the code is below.

```
salary3 <- lm(salary ~ bs(years,degree=1,knots=1986))
plot(years,salary,xlab="Years",ylab="Salary change",pch = 19)
abline(h=0,lty=2)
lines(years,fitted(salary3))
```

The **anova** function can be used to compare the fit of these three models.

```
anova(salary1, salary2, salary3)

Analysis of Variance Table

Model 1: salary ~ bs(years, degree = 1)
Model 2: salary ~ bs(years, degree = 1, df = 2)
Model 3: salary ~ bs(years, degree = 1, knots = 1986)
   Res.Df      RSS Df Sum of Sq      F  Pr(>F)
1      26  229.385
2      25  191.594  1    37.791 4.9311 0.03567 *
3      25  148.323  0    43.271

---
Signif. codes:  0 '***' 0.001 '**' 0.01 '*' 0.05 '.' 0.1 ' ' 1
```

The comparison between the first two models shows a significant improvement, $F(1, 25) = 4.93, p = .04$. There is only one further degree of freedom used because to draw the additional line all you need to know is its slope, since the knot is assumed to be the median of the **years** variable. The third model fits better. The residual sum of squares is lower when we place the knot at 1986 than at the median. There is no significance test for whether this is better than the second model because the two use the same degrees of freedom in their models. In other words, they each estimate the same number of things (the intercept and the slopes of the two lines). Looking at Figure 7.4, the third model appears to fit well and shows that since 1986 the trend is that academics' salary increases have been getting smaller in real terms (as academics, our main hope is that the spline stays above the dashed 0% line).

The next set of models use 2 degree polynomials (i.e., quadratics). Here is the code to create each of these models and to graph them in the bottom row of Figure 7.4. When **bs** connects curves it does so in the smoothest way possible. For these curves it means that they connect in a straight line. This makes the mathematics complex, but it means only one extra degree of freedom is needed for each knot. This is why **df=3** is appropriate for **salary5**.

```
salary4 <- lm(salary ~ bs(years,degree=2,df=2))
plot(years,salary,xlab="Years",ylab="Salary change",pch = 19)
abline(h=0,lty=2)
lines(years,fitted(salary4))
salary5 <- lm(salary ~ bs(years,degree=2,df=3))
plot(years,salary,xlab="Years",ylab="Salary change",pch = 19)
abline(h=0,lty=2)
lines(years,fitted(salary5))
```

```
salary6 <- lm(salary ~ bs(years,degree=2,knots=1986))
plot(years,salary,xlab="Years",ylab="Salary change",pch = 19)
abline(h=0,lty=2)
lines(years,fitted(salary6))
par(mfrow=c(1,1))
```

These models can be compared among themselves[1]:

anova(salary4,salary5,salary6)

```
Analysis of Variance Table

Model 1: salary ~ bs(years, degree = 2, df = 2)
Model 2: salary ~ bs(years, degree = 2, df = 3)
Model 3: salary ~ bs(years, degree = 2, knots = 1986)
  Res.Df     RSS Df Sum of Sq      F Pr(>F)
1     25 177.740
2     24 170.320  1     7.421 1.0457 0.3167
3     24 174.845  0    -4.526
```

and they can be compared with the linear models. For example,

anova(salary2,salary5)

```
Analysis of Variance Table

Model 1: salary ~ bs(years, degree = 1, df = 2)
Model 2: salary ~ bs(years, degree = 2, df = 3)
  Res.Df     RSS Df Sum of Sq      F  Pr(>F)
1     25 191.594
2     24 170.320  1    21.274 2.9978 0.09621 .
---
Signif. codes:  0 '***' 0.001 '**' 0.01 '*' 0.05 '.' 0.1 ' ' 1
```

shows that the quadratic model with one knot at the median (the default location for a single knot in **bs**) is not a significant improvement upon the linear model, $F(1,24) = 3.00, p = .10$. Interestingly, the linear one with the knot at 1986 fits even better than the quadratic one. This is because, as well as having two quadratic curves, it has the requirement to have a smooth transition between the curves, so as Figure 7.4 shows, it does not get very near the outliers at the top of the plot.

Finally, the default spline in the **bs** function is a cubic curve (so **degree=3**) with no knots. So writing **lm(salary~bs(years))** produces a cubic function through the data (the same as **lm(salary~poly(years,3))**). To add a knot, use **lm(salary~bs(years, df=4))**.

---

[1]When comparing models with different splines, allowing more flexibility does not mean that the less complex model is nested within the more complex model. This is because of the way the splines are constrained: different contrasts are used rather than just having additional ones (as with polynomial contrasts). This means the models are not necessarily nested, so care is required using F values to compare models. These difficulties are more apparent with more complex models than are typical in psychology.

Example 10 – Wages with experience, education and gender

* Data: Berndt (1991) and original accessed from the Carnegie Mellon data archive: http://lib.
  stat.cmu.edu/datasets/CPS_85_Wages.
* Package: **splines** and **gam**.
* Research question: After taking into account education level and gender, does experience
  have a linear relationship with the log of wages or a more complicated one? These data are
  often used to answer another question, whether men make more money than women after
  taking into account various variables including education and experience.
* Purpose: To show using splines to allow for a non-linear covariate.

We will go through an example using data from 534 respondents on hourly wages
and several covariates (experience in years, gender and education in years) (Berndt,
1991). There was one outlier with an hourly wage of \$44 ($z = 6.9$) and it was removed,
but the data remained skewed (1.28, $se = 0.11$). Logging these data removes the skew
(0.05, $se = 0.11$), so a fairly common approach is to use the logged values as the response
variable and assume that the residuals are normally distributed. These data, with the
outlier removed, can be accessed in the usual way. The package **splines** is activated
and the variables names are shown.

```
webreg <- "http://www.sagepub.co.uk//wrightandlondon//"
cp85 <- read.table(paste(webreg,"cp85.dat",sep=""), header=T)
library(splines)
attach(cp85)
names(cp85)
```

```
[1] "EDUC" "FEMALE" "EXPER" "WAGE" "AGE" "LNWAGE"
```

Suppose the researchers' main interests are with the experience variable (**EXPER**),
and whether **LNWAGE** steadily increases or whether it increases rapidly until some point
(a knot) and then increases but less rapidly. For argument's sake, let us assume that the
increases are both linear with the logged wages and that placing the knot at the median
of **EXPER** is okay. The researchers accept that wages increase with education (**EDUC**)
and believe that the relationship may be nonlinear, and so allow this to be modeled with
a spline. As the variable **FEMALE** is binary only a single parameter is needed to measure
the difference in earnings between males and females. While categorical variables can
be included within GAMs, the purpose of GAMs is to examine the relationships between
quantitative variables and the response variable. The first model is:

$$\ln wages_i = \beta 0 + \beta 1 female_i + \beta 2 Exper_i + f1(Educ_i) + e_i.$$

This is like a normal multiple linear regression for the variables **FEMALE** and **EXPER**,
the model fits both as conditionally linear with **LNWAGE**, but the relationship for **EDUC**
is more complex.

As with many of the regression procedures discussed in this book we will compare two
models which vary in how much information they use to predict the response variable.
The first one we create will be the simpler. To predict **LNWAGE** we use a single straight
line for **EXPER**. This is shown by using the **bs** function, for B-spline, but telling it
that **degree=1** which means to use straight lines and **df=1**. As explained in the

previous example, **df=x** refers to total degrees of freedom for the spline. Since it is the same as the degree of the polynomial (i.e., **1**), it means there are no knots and only a single line is used. We expect the relationship between **EDUC** and **LNWAGE** to be non-linear. For **EDUC** we use a flexible spline with **degree=3** (cubic) and a single knot (**df=4**). Cubic splines are often used because when they are placed together it is difficult to see where the knots lie so that the curve appears very smooth. This is the default for some of the spline functions (like **ns**), but for simplicity we will continue to use **bs**. The model and some summary information are:

```
wage1 <- lm(LNWAGE~bs(EXPER,df=1,degree=1)
   + bs(EDUC,df=4) + FEMALE)
summary(wage1)


Call:
lm(formula = LNWAGE ~ bs(EXPER, df = 1, degree = 1)
   + bs(EDUC, df = 4) + FEMALE)


Residuals:
     Min        1Q    Median        3Q       Max
-2.144173 -0.295100 -0.002809  0.307351  1.144169


Coefficients:
                        Estimate Std. Error t value Pr(>|t|)
(Intercept)              1.18445    0.34391   3.444 0.000619 ***
bs(EXPER, df = 1,
   degree = 1)           0.73050    0.09550   7.649 9.72e-14 ***
bs(EDUC, df = 4)1        0.02196    0.50041   0.044 0.965009
bs(EDUC, df = 4)2        0.38501    0.32881   1.171 0.242163
bs(EDUC, df = 4)3        1.17094    0.36977   3.167 0.001631 **
bs(EDUC, df = 4)4        1.18195    0.34541   3.422 0.000670 ***
FEMALE                  -0.26411    0.03897  -6.777 3.30e-11 ***
---
Signif. codes:  0 '***' 0.001 '**' 0.01 '*' 0.05 '.' 0.1 ' ' 1


Residual standard error: 0.4443 on 526 degrees of freedom
Multiple R-Squared: 0.2861,      Adjusted R-squared: 0.278
F-statistic: 35.13 on 6 and 526 DF,   p-value: < 2.2e-16
```

Useful information can also be gathered from the **anova** function.

```
anova(wage1)


Analysis of Variance Table


Response: LNWAGE
                        Df   Sum Sq Mean Sq F value    Pr(>F)
bs(EXPER, df = 1,
   degree = 1)           1    2.006   2.006  10.163  0.001518 **
bs(EDUC, df = 4)         4   30.537   7.634  38.678 < 2.2e-16 ***
```

```
FEMALE                        1    9.064    9.064   45.924 3.305e-11 ***
Residuals                   526 103.821    0.197
---
Signif. codes:   0 '***' 0.001 '**' 0.01 '*' 0.05 '.' 0.1 ' ' 1
```

This output tells us that all the variables are significant predictors of **LNWAGE**, and that for **EDUC** this is as a fairly complex curve through the data. As with the previous example much of this output is difficult to interpret without graphs. When there are several predictor variables the methods used in the previous figures need to be adapted. Two methods will be used. In Figure 7.5 the **predict** function is used, as with Figures 7.1–7.4, but split by gender and with different lines for different amounts of experience. In Figure 7.6 the **gam** function (Hastie, 2008) is used. For complex problems it is easier to use this function rather than **lm** (or **glm** for the next example), and it can also be used with other smoothing functions.

The left and right panels of Figure 7.5 show the relationship between **EDUC** and **LNWAGE** for males and females, respectively. Within each panel five different lines are used to show different levels of experience. The five lines correspond to the minimum, first quartile, median, third quartile, and the maximum (basically the 5-point summary, Tukey, 1977). These are found with:

```
experq <- quantile(EXPER)
```

We then use the **predict** function with the model, **wage1**, but providing the function with new values for **EXPER**, **EDUC**, and **FEMALE**. Because **EDUC** goes from 2 to 18, a good range to plot values is from 0 to 20. To create a mini-dataset with these values for each of the five quantiles of males (which have the value 0 for **FEMALE**) is:

```
minimales <-
    data.frame(EXPER=rep(experq,each=21),EDUC=rep(0:20,5),
    FEMALE=rep(0,105))
```

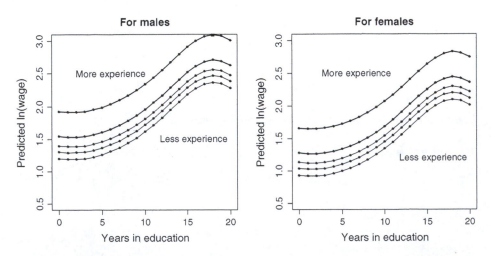

Figure 7.5   The predicted values from **wage 1** for different amounts of experience for males and females

and for females is:

```
minifemales <-
    data.frame(EXPER=rep(experq,each=21),EDUC=rep(0:20,5),
    FEMALE=rep(1,105))
```

The plots and lines can be made as follows. For males:

```
par(mfrow=c(1,2))
    plot(minimales$EDUC,predict(wage1,minimales),
    xlab="Years in education", ylab="Predicted ln(wage)",
    pch=19,cex=.5,ylim=c(.5,3))
lines(0:20, predict(wage1,minimales[1:21,]),lwd=.5)
lines(0:20, predict(wage1,minimales[22:42,]),lwd=1)
lines(0:20, predict(wage1,minimales[43:63,]),lwd=1.5)
lines(0:20, predict(wage1,minimales[64:84,]),lwd=2)
lines(0:20, predict(wage1,minimales[85:105,]),lwd=2.5)
text(6,2.5,"More experience")
text(15,1.5,"Less experience")
title("For males")
```

and for females:

```
plot(minifemales$EDUC,predict(wage1,minifemales),
    xlab="Years in education", ylab="Predicted ln(wage)",
    pch=19,cex=.5,ylim=c(.5,3))
lines(0:20, predict(wage1,minifemales[1:21,]),lwd=.5)
lines(0:20, predict(wage1,minifemales[22:42,]),lwd=1)
lines(0:20, predict(wage1,minifemales[43:63,]),lwd=1.5)
lines(0:20, predict(wage1,minifemales[64:84,]),lwd=2)
lines(0:20, predict(wage1,minifemales[85:105,]),lwd=2.5)
text(6,2.5,"More experience")
text(15,1.2,"Less experience")
title("For females")
par(mfrow=c(1,1))
```

Making Figure 7.5 is a hassle and easily open to error. If your interest is just in the shape of the spline for education, an easier method is using the **gam** function. After loading the **gam** package, you run the model, but with the **gam** function rather than **lm**:

```
install.packages("gam")
library(gam)
wage1a <- gam(LNWAGE~bs(EXPER,df=1,degree=1) + bs(EDUC,
    df=4) + FEMALE)
```

The **summary** function shows that this is the same model as **wage1**, as shown by the output from **anova(wage1)**. The residual deviance of **wage1a** (103.8207) is the same as the residual sum of squares from **wage1** (103.821), and the difference between the null deviance and residual deviance for **wage1a**

(145.4278 − 103.8207 = 41.6071) is the same as the total model sum of squares from **wage1**: 2.006 + 30.537 + 9.064 = 41.607.

```
summary(wage1a)

Call: gam(formula = LNWAGE ~ bs(EXPER, df = 1, degree = 1)
  + bs(EDUC,
      df = 4) + FEMALE)

Deviance Residuals:
      Min         1Q     Median         3Q        Max
-2.144173  -0.295100  -0.002809   0.307351   1.144169

(Dispersion Parameter for gaussian family taken to be 0.1974)

    Null Deviance: 145.4278 on 532 degrees of freedom
Residual Deviance: 103.8207 on 526 degrees of freedom
AIC: 656.6771

Number of Local Scoring Iterations: 2
DF for Terms


                                  Df
(Intercept)                        1
bs(EXPER, df = 1, degree = 1)      1
bs(EDUC, df = 4)                   4
FEMALE                             1
```

gam is for generalized additive models, and as with the generalized linear models discussed in Chapter 6, a dispersion parameter is estimated. It is the null deviance divided by the residual degrees of freedom (103.8207/526). The reason why we use **gam** here is because when a **gam.object** is used in **plot**, it shows the shapes of all the predictor variables. Figure 7.6 shows these. The standard error intervals are shown with dashed lines by using the option **se=T**. The code is much shorter than that used constructing Figure 7.5.

```
par(mfrow=c(1,3))
plot(wage1a, se=T)
```

Figure 7.5 shows that as **EXPER** increases so does **LNWAGE**. It shows that as **EDUC** increases **LNWAGE** also increases. Finally, as gender moves from male to female, **LNWAGE** decreases. Of course, with a variable like **GENDER** this does not make sense. If **FEMALE** was stored as a factor then boxplots would be shown in this final graph. Try re-running the analyses with **FEMALE2 <- as.factor(FEMALE)** in the **gam** command that created **wage1a**.

Next, a slightly more complicated spline is used for **EXPER**. Two straight lines, connected at a knot at the median of **EXPER**, are used. This can be run either with **lm** or **gam**. We will use **gam** since the resulting graphs are easier to construct.

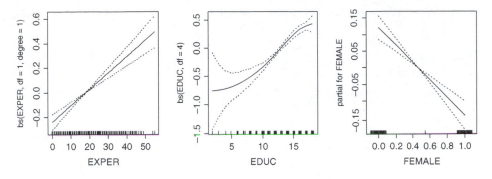

Figure 7.6 The graphs made for the generalized additive model on the log of wages

To change this, the option **df=1** should be changed to **df=2** with the **bs** function for **EXPER**:

```
wage2a <- gam(LNWAGE~bs(EXPER,df=2,degree=1) + bs(EDUC,
    df=4) + FEMALE)
```

This model can be compared to the previous one with the **anova** function. To make consistent with the previous comparison the **F** test option is used.

```
anova(wage1a,wage2a,test="F")

Analysis of Deviance Table

Model 1: LNWAGE ~ bs(EXPER, df = 1, degree = 1) + bs(EDUC, df = 4)
    + FEMALE
Model 2: LNWAGE ~ bs(EXPER, df = 2, degree = 1) + bs(EDUC, df = 4)
    + FEMALE
  Resid. Df Resid. Dev  Df Deviance      F   Pr(>F)
1       526    103.821
2       525     97.689   1    6.131 32.952 1.599e-08 ***
---
Signif. codes:  0 '***' 0.001 '**' 0.01 '*' 0.05 '.' 0.1 ' ' 1
```

This allows us to say that the model with a single knot fits significantly better than the model without a knot, $F(1,525) = 32.95$, $p < .01$. The AICs of the two models can also be compared by looking at the output from the **summary** functions; AIC drops from 656.68 to 626.23. Remember, AIC and BIC are measures of the residual – the larger their values the worse the fit. Similarly, the proportion of total deviance accounted for by the model increases from $(145.4278 − 103.8207)/145.4278 = .286$ for **wage1a** to $(145.4278 − 97.6892)/145.4278 = .328$ for **wage2a**.

To understand the model better it should be graphed. Figure 7.7 shows the results of:

```
plot(wage2a,se=T)
```

These show the same basic pattern for **EDUC** and **FEMALE**. The difference is for **EXPER**. This shows that the log of salary increases rapidly at the beginning, but less so afterwards.

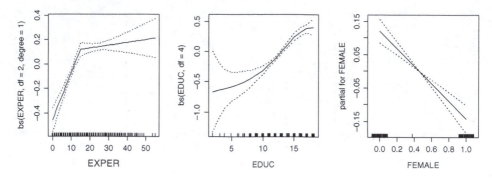

Figure 7.7   The plots from the generalized additive model of `wage2a`

A caveat is worth adding here, but it applies to most statistical models. It is important not to interpret this result to mean that the model with one knot is right. Models after all are just models of reality, and the usual interpretation of this means that they are never 'right,' but they may be very good. Cox (2006: 31) describes this well: 'the very word "model" implies idealization. With very few possible exceptions it would be absurd to think that a mathematical model is an exact representation of a real system.'

Finally, many readers will be interested in the gender gap in pay. Because an argument for gender inequality in the past has been that it was due to experience and education, it is worth using very flexible splines for these variables and seeing whether **FEMALE** is still significant.

```
summary(lm(LNWAGE ~ bs(EXPER, degree=3,
    df=4)+bs(EDUC,degree=3, df=4) + FEMALE))
```

```
Call:

lm(formula = LNWAGE ~ bs(EXPER, degree = 3, df = 4) + bs(EDUC,
    degree = 3, df = 4) + FEMALE)

Residuals:
     Min       1Q   Median       3Q      Max
-2.20144 -0.27374  0.02893  0.28891  1.19476

Coefficients:
```

	Estimate	Std. Error	t value	Pr(>|t|)	
(Intercept)	0.99061	0.37215	2.662	0.008010	**
bs(EXPER, degree = 3, df = 4)1	0.53593	0.14628	3.664	0.000274	***
bs(EXPER, degree = 3, df = 4)2	0.82210	0.15968	5.148	3.73e-07	***
bs(EXPER, degree = 3, df = 4)3	0.82448	0.25049	3.291	0.001064	**
bs(EXPER, degree = 3, df = 4)4	0.72980	0.28376	2.572	0.010390	*
bs(EDUC, degree = 3, df = 4)1	-0.02853	0.50956	-0.056	0.955373	
bs(EDUC, degree = 3, df = 4)2	0.15374	0.36227	0.424	0.671463	
bs(EDUC, degree = 3, df = 4)3	1.00241	0.39159	2.560	0.010753	*

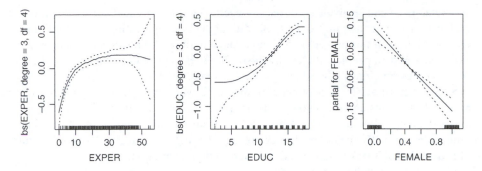

Figure 7.8   The graphical output with cubic splines with a single know each for **EXPER** and **EDUC**. This shows, even allowing for very flexible main effects between these variables and **LNWAGE**, that the gender still makes a difference

```
bs(EDUC, degree = 3,
  df = 4)4                      0.95806    0.37396    2.562 0.010687 *
FEMALE                        -0.26329    0.03778   -6.970 9.62e-12 ***
---
Signif. codes:  0 '***' 0.001 '**' 0.01 '*' 0.05 '.' 0.1 ' ' 1

Residual standard error: 0.4304 on 523 degrees of freedom
Multiple R-Squared: 0.3337,     Adjusted R-squared: 0.3222
F-statistic:  29.1 on 9 and 523 DF,  p-value: < 2.2e-16
```

So, the difference still remains, $t(523) = 6.97, p < .001$. Figure 7.8 shows the graphical output:

```
plot(gam(LNWAGE~bs(EXPER,degree=3,df=4)+
    bs(EDUC,degree=3,df=4)+ FEMALE),se=T)
par(mfrow=c(1,1))
```

Example 11 – Detecting deception

- Data: Hypothetical, but capturing the flavor of Vrij (2005).
- Research questions: Are criteria-based content analysis (CBCA) scores equally diagnostic of truth for children of different ages? Often this would be done with the assumption that everything was linearly related. Here we do not wish to make this assumption.
- Purpose: To illustrate splines/GAMs with a binary response variable.

In Chapter 6 the generalized linear model was shown to be an extension of the linear model. Additive models using splines can also be used with generalized models. The **glm** function or the **gam** function can be used with B-splines using **bs**. The **glm** function provides more useful numeric output and the **gam** function is easier to use for graphs. Further, the **gam** function is necessary for complex splines not covered in this book. In this example the focus will be on graphs, so the **gam** function will be used.

To illustrate a logistic additive model, data inspired by truth and lie detection using criteria-based content analysis (CBCA) (Vrij, 2005) will be used. This is a method used in several countries to try to determine whether a child is telling the truth or a lie when questioned, usually in connection with cases of child sexual abuse. There are 19 criteria and a statement can be given a 0, 1, or 2 for each criterion, and these are summed so that each person can get a score from 0 to 38 with high scores indicating more truthfulness. One problem with this procedure is that people with more linguistic skills tend to have higher scores than people with less linguistic skills. Because of this there is assumed to be a complex relationship between age, CBCA score, and truth.

Suppose there are 1000 statements from people who are 3–22 years old, and these have CBCA scores and it is known whether the statements are truthful or not. We will create the data below, so that we know age and truth should be independent (and in the sample they are). Three GAMs were estimated. The first has just CBCA to predict truth. This uses the logit link function and assumes binomial variation.

We created the three variables here. Notice in calculating **truth**, **age** is not used, thus the two are independent in one sense. The coding for **cbca** looks complex. We were playing with the coding to make it look like what is suggested in Vrij's review. Nobody has done a deception study with 1000 people.

```
set.seed(406)
age <- runif(1000,3,23)
truth <- rbinom(1000,1,.5)
cbca <-round(-3*truth+2*truth*log(age)
    +.2*age+rbinom(1000,25,.7))
```

The following shows that there is not a statistical significant relationship between **age** and **truth**. This is the basic logistic regression (see Chapter 6).

```
summary(glm(truth~age,family=binomial))

Call:

glm(formula = truth ~ age, family = binomial)

Deviance Residuals:
    Min      1Q   Median      3Q      Max
 -1.214  -1.189    1.142   1.165    1.190

Coefficients:
              Estimate  Std. Error  z value  Pr(>|z|)
(Intercept)   0.102594    0.154826    0.663     0.508
age          -0.005776    0.010941   -0.528     0.598

(Dispersion parameter for binomial family taken to be 1)

    Null deviance: 1386.1  on 999  degrees of freedom
Residual deviance: 1385.8  on 998  degrees of freedom
AIC:                1389.8

Number of Fisher Scoring iterations: 3
```

It is worth making sure that there was not an odd relationship between **truth** and **age,** for example younger and older children being truthful, but those in the middle lying.

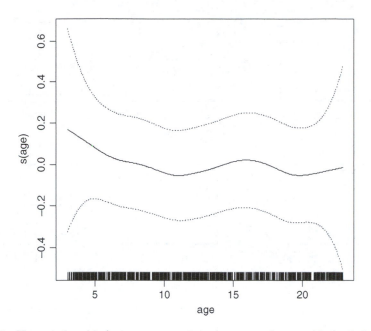

Figure 7.9   The relationship between **truth** and **age** as shown with the default smoothing spline. There appears no relationship

Figure 7.9 shows the graph for this relationship using **gam** and **bs**. The spline is two cubic curves (**degree=3** is the default) connected at a single knot at the median of age (the median being the default). By setting **se=T** R adds lines for $\pm se$. The end points on splines tend to have large confidence intervals. There are some methods to adjust for this (see Wood, 2006). Because this is a logistic model we have written **family=binomial** (see Chapter 6).

```
library(gam)
plot(gam(truth~bs(age,df=4),family=binomial),se=T)
```

Now we focus on the relationship between **cbca** and **truth**. The **bs(cbca, df=4)** in the function below means the spline is two cubic curves connected at the median. For most psychology examples this provides enough flexibility, although for large data set you may wish to increase the number of knots by increasing **df=** to a higher number.

```
gam0 <- gam(truth~bs(cbca, df=4), family=binomial)
```

The top left panel of Figure 7.10 shows that there is a positive association between **cbca** and **truth**. If you are making a graph using the **predict** function with the **glm** function, use **type="response"** as in Chapter 6, if you want the y-axis to be in the original scale. Next the **age** variable is added. We have used the same flexibility in the spline for this variable.

```
gam1 <- gam(truth ~ bs(cbca, df=4) + bs(age, df=4),
    family=binomial)
```

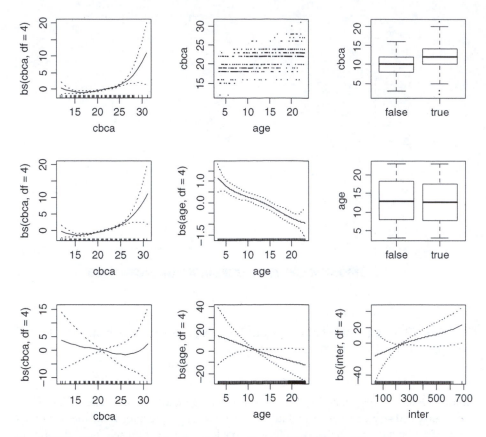

Figure 7.10 The basis function for 3 generalized additive models for the (made-up) CBCA data. The first row shows the spline between **cbca** and **truth** (for **gam0**); the scatterplot of **cbca** with **age**, and boxplots showing truthful statements had, overall, higher **cbca** scores. The second row shows the splines between **cbca** and **truth**, and **age** and **truth** (for **gam1**); and boxplots showing that the distributions of **age** for false and true statements are similar. The third row shows the splines for **gam2**. The effect of most interest is the interaction. It shows that the diagnostic value of CBCA scores increases with children's age

The two models can be compared using the **anova** function. For these generalized models, **test="Chisq"**, which is the default, is appropriate.

```
anova(gam0,gam1)

Analysis of Deviance Table

Model 1: truth ~ bs(cbca, df = 4)
Model 2: truth ~ bs(cbca, df = 4) + bs(age, df = 4)
  Resid. Df Resid. Dev  Df Deviance P(>|Chi|)
1       995    1271.29
2       991    1222.43   4    48.86 6.256e-10
```

This shows including **age** increases the predictive value for predicting **truth**, $\chi^2(4) =$ 48.86, $p < .001$. The second row of Figure 7.10 shows that, once controlling for **cbca**, as **age** increases the likelihood of the statements being truthful decreases. This conditional relationship occurs because **cbca** and **age** are positively correlated (see second panel, top row, of Figure 7.10).

When using some types of splines, and with other complex models, the number of degrees of freedom will not be a whole number. Degrees of freedom is just a measure of information so non-integer values present no conceptual difficulties. With the models generated by **bs** the degrees of freedom should be whole numbers (or at least within rounding error).

The main hypothesis, that the interaction improves the fit of the model, is now included. There was an interaction term used in creating the data, so we would expect this to be significant. If the effect were non-significant it would be a Type 2 error. We created the interaction, **inter <- cbca*age**, prior to running the **gam** function.

```
inter <- cbca*age
gam2 <- gam(truth~bs(cbca,df=4)+bs(age,df=4)+bs(inter,
    df=4), family=binomial)
```

When this model is compared with the previous,

```
anova(gam1,gam2)

Analysis of Deviance Table

Model 1: truth ~ bs(cbca, df = 4) + bs(age, df = 4)
Model 2: truth ~ bs(cbca, df = 4) + bs(age, df = 4) +
    bs(cbca * age, df = 4)

  Resid. Df Resid. Dev  Df Deviance P(>|Chi|)
1       991    1222.43
2       987    1197.76   4    24.67 5.863e-05
```

this shows that the interaction is significant: $\chi^2(4) = 24.67$, $p < .001$. It is best to interpret this while looking at the graphs.

Figure 7.10 is a 3×3 panel graph that illustrates the models in this example. We begin by telling R to treat **truth2** as a categorical variable. This means in the **plot** commands below it draws boxplots rather than scatterplots when **truth2** is on the x axis. The **se=T** adds the standard error lines. The **pch="."** tells R to make the symbols for the scatterplot as small as possible. This is often necessary when you have large amounts of data.

```
par(mfrow=c(3,3))
truth2 <- factor(truth,label=c("false","true"))
plot(gam0, se=T)
```

```
plot(age,cbca,pch=".")
plot(truth2,cbca,ylab="cbca")
plot(gam1, se=T)
plot(truth2,age,ylab="age")
plot(gam2, se = T)
par(mfrow=c(1,1))
```

The first row shows that there is a positive bivariate relationship between **cbca** and **truth**. The relationship looks like the probability of a statement being truthful goes up with **cbca**, mostly for large values of **cbca**, but that at lower values **cbca** is not very diagnostic. The second panel in the first row shows **age** and **cbca** are positively associated and the third panel shows that true statements tend to have higher **cbca** scores than false ones. The second row shows the relationship between **cbca** and **truth** continues after controlling for **age**. **age** has a strong negative relationship with **truth**, but it is important to recall that this is after controlling for **cbca**. As the boxplots to the right show, there is not a bivariate relationship between **age** and **truth**. The third row shows the interaction model. The interaction is shown in the final panel of this row and it is clear that the diagnostic value for **cbca** increases linearly with **age**. In other words, the **cbca** is only valid for older children (according to these made-up data!).

KERNEL METHODS

There are several smoothing functions, other than B-splines, which are used by researchers. One popular type is called kernel methods. These calculate a line for each point along the *x* axis and join these together. The line is based on the data within some distance of that point on the *x* axis. The smoothness of the curve is dependent on how narrow or wide this region (or kernel) is. The wider the region is the smoother the curve. These methods are often called locally weighted methods because only values near that point on the *x* axis are used in calculating the line. Sometimes the functions used will weight all the values in the region equally; sometimes their weight will be dependent on how far they are from the *x* value. The procedures can use ordinary least squares or robust (see Chapter 9) methods. In R the functions **lo**, **loess**, and **lowess** are the most common. Kernel methods can be mathematically complex.

REVISITING LONG-TERM MEMORY

In Chapters 3 and 4 we examined different ways of analyzing data of children's memory (London et al., in press). Children were shown a magic show and asked to recall what they could remember, both soon after the event and at a final interview approximately one year later. The interest was in age differences with respect to changes in the amount reported. In Chapter 4 we described how an ANCOVA is preferred to an ANOVA on the differences. The latter approach assumes memory does not decline over time (see Wainer, 1991). We noted two further considerations in the analysis which we can now address

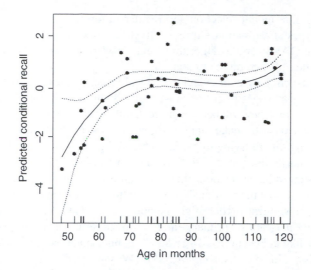

Figure 7.11   The plot for a generalized additive model with a *df* = 4 *B*-spline for age in months predicting final recall partialling out initial recall. Poisson error is assumed with the ln link function

with the tools covered in Chapters 6 and 7. We noted that there was the problem of floor effects: that somebody who only recalled 1 or 2 items could not decline as much as someone who initially recalled 8 or 9 items. While floor effects can never be completely resolved with statistical tools, a Poisson regression works well for count data like these which have a large skew. In addition, there was worry about whether the relationship between age in months and the final recall, after accounting for initial recall, really was linear. It is more likely that memory skills increase rapidly in earlier years and then level off. Now with GAMs as part of our statistical arsenal we can further explore these data. Several models were explored, but the model shown in Figure 7.11 fits the data well. The following code produces this graph. As expected, memory retention increases rapidly for young children, but this level of improvement does not carry on as the children get older.

```
attach(lordex)
gam1 <- gam(Final ~ Initial + bs(AGEMOS,df=4),poisson)
plot(gam1,se=T,terms="bs(AGEMOS, df = 4)",1,pch=19,
    xlab="Age in months",ylab="Predicted conditional
    recall",cex.lab=1.5,cex.axis=1.3)
```

SUMMARY

GAMs are useful generalizations of the basic regression models. Their basic building blocks are the splines used to model the relationships. Like GLMs, they allow different link functions and distributions which are appropriate for a large amount of data collected in psychology and other sciences. Further, the additive components allow an extremely flexible approach to data modelling. There are several extensions to GAMs not discussed here, like model selection and regularization techniques, multilevel GAMs, and different types of estimation. Current software allows many different types of curves to be fit within GAMs, and as algorithms and software advance, these models should become more flexible and more popular.

The typical psychology dataset and the typical research questions within psychology dictate how splines/GAMs might be most useful within our discipline. There are three ways in which we feel they will become more popular. The first is in an ANCOVA context when you do not wish to make many assumptions about the relationship between the covariate, which you are trying to partial out, and the response variable. In these cases you are not primarily interested in the covariate, but in the other predictors, and you do not want to assume a linear relationship between the covariate and the response variable. If you knew more about the relationship between the covariate and the response variable, then a specific non-linear model might be appropriate, but often you do not have this knowledge (and it is not of primary interest). In these cases we recommend a fairly flexible spline, like cubic curves (**degree=3**) connected at one or two knots (so **df=4** or **df=5**). This could be called a GAMCOVA (Wright, 2008). The second way is where you want to test if the data are consistent with a linear model, or whether some more flexible model is better. In many undergraduate courses people are taught to try a quadratic or a cubic term, and these methods are one alternative, but the flexibility of splines make them an attractive option. Again, **degree=3** is recommended and usually with just a single knot (**df=4**). The examples presented in this chapter illustrate these uses. The third way that these models are often used is graphically to draw a curve through data. How much flexibility you have is dependent on how much data you have and how smooth you want the curve. Sometimes the kernel methods described in the box above can be used.

SOME WORDS/CONCEPTS WORTH REMEMBERING

R concepts

* plotting with the **predict** function including different levels of other variables.

R functions

* **axes=F** with **box()**: to make a box around data without axes;
* **poly**: to make polynomials of different degrees;
* **degree**: the option for the degree polynomial;
* **predict**: for making predicted values from regressions;
* **bs**: for making B-splines;
* **df**: the option for the spline complexity;
* **knots**: the option for number or location of knots;
* **gam**: for generalized additive models.

Statistical concepts

* polynomial regression: using different degree polynomials;
* piecewise polynomials: adding together polynomials;
* splines: adding them together at knots;
* generalized additive model: models with splines (and other functions).

FURTHER READING

Most of the literature tends to be written for those with much statistical background. Two of the best of these are:

Hastie, T. & Tibshirani, R. (1990) *Generalized additive models*. London: Chapman and Hall. The original GAM book and the basis for the **gam** software. It is more technical than is appropriate for most psychologists.

Wood, S. N. (2006) *Generalized additive models: An introduction with R*. Boca Raton, F.L: Chapman & Hall/CRC. Simon Wood has written an extension of the **gam** package used here. It is a good alternative to **gam**, and is very similar in its use. This is a good book, although a bit technical for most psychologists. http://www.maths.bath.ac.uk/~sw283/

A very introductory piece is:

Wright, D. B. (2008) A new improved analysis of covariance. *Psychologist, 21*, 225–226.

8

Multilevel models

Learning outcomes

1. To be able to run multilevel linear and generalized multilevel linear models.
2. To understand why ignoring clustering in variables is wrong.
3. To have some practice aggregating variables.

In almost every undergraduate methods course students are told that their statistical models assume that the data are independent, but they are not told what to do if the data are not independent. This is because, until about 25 years ago, there were neither the algorithms nor the computational power for conducting many of the statistics that can now be estimated quickly on a desktop computer. Before this, there were corrections that could be applied to non-independent data, but these were cumbersome, problematic, and inflexible.

A common situation is where the non-independence is due to the data being clustered. Multilevel modeling (often called hierarchical modeling) takes into account the clustering. It allows modeling to be conducted simultaneously at the level of the cluster and at the level of the individual. The model does make assumptions. For example, most multilevel models assume that after taking into account the clustering (and any other variables in the model) that the data are independent. Further, for inference the researcher usually assumes that the units at one level are a random sample of all those from within the cluster.

Consider the following example from Barth and colleagues (2004). They looked at the peer relationships of about 600 pupils by the pupils' race and gender. The pupils were sampled from 65 different classrooms. The assumption is that the pupils sampled were representative of the pupils within these classrooms. For illustration, consider their fourth grade sample (age around 10 years old). For the traditional single level approach, a researcher might use a regression of the following form:

$$Peer_i = \beta 0 + \beta 1\, Gender_i + \beta 2\, Race_i + e_i$$

where high $Peer_i$ values mean problematic relations, $Gender_i$ is a dummy variable with female $= 0$ and male $= 1$, and $Race_i$ is a dummy variable with Caucasian $= 0$ and non-Caucasian $= 1$ (in R this would be **lm(Peer~Gender+Race)**). The standard regression assumes that the e_i are independent. Here they are not because there are likely to be classroom effects: children within the same classroom are more likely to have similar $Peer_i$ scores than children from different classrooms. In fact, in Barth et al. about 12% of the total variation in $Peer_i$ could be attributable to classroom level variation. The standard errors using this traditional approach are likely to be too small which means that you will often get a significant p value when you should not. Because in psychology people worry more about Type 1 errors than Type 2 errors (for better or worse), this bias causes much concern among editors, reviewers, and supervisors, who now often require authors to use multilevel modeling.[1]

For multilevel modeling, let the intercepts vary randomly for each classroom. In notation, let the intercepts be $\beta 0_j = \beta 0 + u_j$, where the u_j are independent and normally distributed for the schools. u_j is a residual or error term but at the cluster level. The subscript j is for the 65 schools. Barth et al. (2004) estimated the following model:

$$Peer_{ij} = \beta 0 + \beta 1\, Gender_{ij} + \beta 2\, Race_{ij} + u_j + e_{ij}$$

They found a significant gender effect ($t(1325) = 4.42$) and a significant effect for race ($t(1325) = 5.80$) with males and non-Caucasian participants having poorer peer relations after taking into account each other.

Multilevel modeling can be extended in many ways. First, while it is true that the pupils in Barth et al.'s study were nested within classrooms, the classrooms were nested within schools. In their paper, they looked at this using a 3 level model. In addition, variables that are about the classroom (like characteristics of the teacher) and school (like the neighborhood's affluence) could be included, and in their paper some of these were included. The random part of the model can also be made to incorporate more aspects of the data. It is common to see if the slopes also vary among the classrooms. For example, to allow the gender effect to vary among schools, let $\beta 1_j = \beta 1 + v_j$, where the v_j represent the spread around the central gender effect, $\beta 1$.

A frequently asked question about multilevel modeling is what types of data can be used with it. The textbook example is of pupils sampled within classrooms. This seems to satisfy the criterion that we can imagine pupils as a random sample of all those in the classroom. However, multilevel models are now often used for any hierarchical data set regardless of whether the lower level units can be easily thought of as a random sample of some population. Sometimes researchers apply multilevel modeling to a hierarchical data set without thinking whether the data meet all the assumptions of multilevel modeling. There seems to be a belief that using a modern mathematically complex statistic can rectify any methodological difficulty.

Two examples are used in this chapter. The first has the traditional structure, with pupils nested within classrooms. We examine exercise of children nested with classrooms across several conditions (Hill et al., 2007). We use a multilevel equivalent to ANOVA, and in relation to Lord's Paradox and mediation effects (see Chapter 4). The second example has measurements nested within person. This is the most common example within medicine and is increasingly used in psychology. Multilevel modeling is rapidly become the preferred method for repeated measures data. The example is of people's

---

[1] The p values can be too high too, particularly when multilevel models are used to analyze within subject designs.

memory for own and other race faces (Wright et al., 2003). The response variable is binary, whether the participant says they have seen the face before or not. Thus, this is an example of a generalized linear multilevel model.

Example 12 – Getting children to exercise

- Data: Hill et al. (2007).
- Packages: `nlme`, `lattice`.
- Research question: Does an intervention designed to make children exercise work?
- Purpose: To introduce linear multilevel modeling.

The format for this example follows Wright (1998) where four approaches are used to illustrate the strength of multilevel modeling. First, the clustering is ignored (which is wrong, but useful for comparison). Second, the analysis is done at the cluster level on aggregate measures. This approach addresses a different question than analysis of the lower level, and in most cases falls foul of the ecological fallacy (Robinson, 1950), and is wrong. The third approach treats the cluster as a fixed effect and covaries it out. This is a limited and cumbersome approach. The final approach is multilevel modeling.

The purpose of this research was to examine whether providing children with a leaflet based on the 'theory of planned behavior' increases children's exercise (Hill et al., 2007). Five hundred and three children from 22 different classrooms were sampled. Because it would not have been practical to have children in the same classrooms in different conditions, the 22 classrooms were randomly assigned to 4 different conditions (control, and 3 with leaflets). Children were asked the following question before and after the intervention:

> On average over the last three weeks, I have exercised energetically for at least 30 minutes _____ times per week.

Here we will concentrate on the post-intervention scores. The original exercise variable was skewed (0.83). We have not calculated a standard error or confidence intervals for the skewness because these would assume that the data were not clustered. To lessen the skew .5 was added to each value and the square root was taken (new skewness $= -0.10$). This transformed variable will be used.

The data are read directly from the book's web page. The file has a numeric variable, **wcond**, with the values 1–4. This is changed to a factor and given labels. **"L+Quiz"** means participants received the leaflet and a quiz and **"L+Plan"** means participants received the leaflet and made an exercise plan. The library **nlme** (Pinheiro et al., 2008) is loaded. This library has traditionally been the most used within R to conduct multilevel modeling, although the package used in the next example is seen as an extension of it and is likely to become more popular. This library **nlme** may need to be installed first.

```
webreg <- "http://www.sagepub.co.uk//wrightandlondon//"
exer <- read.table(paste(webreg,"exercise.dat",sep=""),
  header=T)
attach(exer)
cond <- factor(wcond, labels=c("Control","Leaftet",
  "L+quiz","L+plan"))
```

```
install.packages("nlme")
library("nlme")
```

The first model is the ordinary least squares regression, which is the equivalent to a oneway anova, so **aov(sqw2~cond)** produces the same model. The intercept is 1.65, which is the predicted value for the control group. All the other coefficients are positive, which shows that the means are all higher for those given a leaflet (alone, with a quiz, or with a plan). Everything looks significant (or almost significant), but the standard errors and *p* values are wrong. Imagine if there happened to be one really sporty class because of a particular teacher. Whichever condition that class was in might have 20 high exercise people.

```
model1 <- lm(sqw2 ~ cond)
summary(model1)

Call:
lm(formula = sqw2 ~ cond)

Residuals:
     Min        1Q    Median        3Q       Max
-1.19166  -0.31763   0.03136   0.33913   1.45818

Coefficients:
             Estimate Std. Error t value Pr(>|t|)
(Intercept)   1.64631    0.04925  33.425  < 2e-16 ***
condLeaftet   0.19316    0.06979   2.768 0.005857 **
condL+quiz    0.13588    0.06926   1.962 0.050318 .
condL+plan    0.25246    0.07127   3.542 0.000434 ***
---
Signif. codes:  0 '***' 0.001 '**' 0.01 '*' 0.05 '.' 0.1 ' ' 1

Residual standard error: 0.5572 on 499 degrees of freedom
Multiple R-Squared: 0.02728,     Adjusted R-squared: 0.02143
F-statistic: 4.665 on 3 and 499 DF,  p-value: 0.003168
```

Figure 8.1 is made in two parts. The first is here, the basic boxplot. The default for **plot** when the first variable is a factor is to draw a boxplot. We used **expression(sqrt (exercise + .5))** to put the mathematical formula along the *y* axis. The function **expression** allows this to be done, and there are a variety of functions that can be printed (type **demo(plotmath)** and see Murrell, 2006: 97). The second part of the graph is the dots for each classroom. These are made in a couple of pages.

```
plot(cond,sqw2,xlab="Conditions",ylab=expression
    (sqrt(exercise + .5)))
```

The second approach to these data involves calculating aggregate variables for the classrooms. The **aggregate** function is used. Its first argument is the variable to aggregate (**sqw1**), the second a **list** of the variable to aggregate by (**class**), and finally the function used to summarize the group (here, the **mean**). The mean for

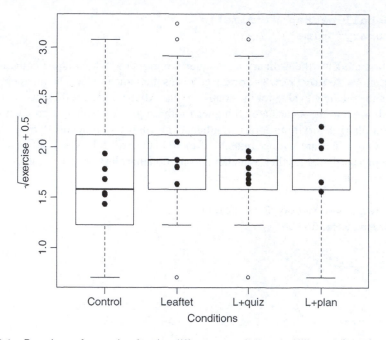

Figure 8.1   Boxplots of exercise for the different conditions in Hill et al. (2007). The large dots are the means for the 22 classrooms

exercise (**mexer**) is calculated for each class, and the same procedure is used to get a variable of the same length corresponding to the condition (**mean** is used, but all values within a class are the same). The **[,2]** tells R to store the **mean** of these in the variables **mexer** and **mcond**. If **[,1]** were used it would have stored the classroom number.

```
mexer <- aggregate(sqw2,list(class),mean)[,2]
mcond <- aggregate(wcond,list(class),mean)[,2]
```

The linear regression (**model2**) looks at the means for the classrooms. Although $R^2 = .22$ is fairly big, the effect is non-significant because the sample size is small (because the sample size is now the number of classrooms). Further, the effect size does not correspond to an effect for how the intervention increases a child's exercise. Making this conclusion would be the ecological fallacy (Robinson, 1950). It is about classrooms (and if the class sizes were a lot bigger the correlation would likely increase, what in psychology is known as the Spearman-Brown prophecy).

```
model2 <- lm(mexer~factor(mcond))
summary(model2)

Call:
lm(formula = mexer ~ factor(mcond))

Residuals:
      Min        1Q     Median        3Q       Max
-0.335601 -0.118965 -0.006939  0.128987  0.313912
```

```
Coefficients:
                 Estimate Std. Error t value Pr(>|t|)
(Intercept)       1.64890    0.07778  21.201 3.51e-14 ***
factor(mcond)2    0.18441    0.11536   1.599   0.1273
factor(mcond)3    0.13449    0.10999   1.223   0.2372
factor(mcond)4    0.24774    0.11536   2.148   0.0456 *
---
Signif. codes:  0 '***' 0.001 '**' 0.01 '*' 0.05 '.' 0.1 ' ' 1


Residual standard error: 0.1905 on 18 degrees of freedom
Multiple R-Squared: 0.2208,      Adjusted R-squared: 0.09092
F-statistic:   1.7 on 3 and 18 DF,  p-value: 0.2027
```

Now that the aggregate variables have been created the points can be added to Figure 8.1, using the **points** command. The points are made fairly big by **cex=1.3**.

```
points(mcond,mexer,pch=19,cex=1.3)
```

The third method is to treat **class** as a covariate, partialling out its effects like you would with an ANCOVA. For some multilevel datasets this is an acceptable alternative, providing the number of classrooms is not large. Here it does not work because each classroom is assigned to a condition, so classroom and condition are confounded. This is why the coefficients for **cond** are not estimated below. This can happen in other situations also, particularly when there are lots of clusters and few points per cluster. Wright (1998) describes how this can be a valid approach to modeling, but that it requires estimating lots of things (and therefore can overfit the model) and from a philosophical standpoint it is difficult to make inference about population values. Both of these disadvantages are addressed by the multilevel approach.

```
model3 <- lm(sqw2 ~ factor(class) + cond)
summary(model3)

Call:
lm(formula = sqw2 ~ factor(class) + cond)

Residuals:
     Min       1Q   Median       3Q      Max
-1.50345 -0.33972  0.04162  0.33896  1.59050

Coefficients: (3 not defined because of singularities)
                 Estimate Std. Error t value Pr(>|t|)
(Intercept)       1.53952    0.11497  13.390  < 2e-16 ***
factor(class)2    0.51270    0.15489   3.310  0.00100 **
factor(class)3    0.25450    0.15764   1.614  0.10710
factor(class)4    0.35974    0.16260   2.212  0.02741 *
factor(class)5    0.24275    0.16452   1.475  0.14074
factor(class)6    0.52554    0.16260   3.232  0.00131 **
factor(class)7    0.14349    0.15764   0.910  0.36317
```

```
factor(class)8      0.26334      0.15489     1.700   0.08974 .
factor(class)9     -0.01277      0.16452    -0.078   0.93815
factor(class)10     0.33220      0.15764     2.107   0.03561 *
factor(class)11     0.02152      0.15764     0.137   0.89145
factor(class)12     0.42091      0.16661     2.526   0.01185 *
factor(class)13     0.67104      0.15622     4.295 2.11e-05 ***
factor(class)14     0.45717      0.16260     2.812   0.00513 **
factor(class)15     0.39234      0.16661     2.355   0.01893 *
factor(class)16     0.18404      0.16889     1.090   0.27640
factor(class)17     0.26929      0.15764     1.708   0.08824 .
factor(class)18     0.13972      0.16082     0.869   0.38539
factor(class)19     0.11036      0.16260     0.679   0.49764
factor(class)20    -0.10572      0.16452    -0.643   0.52078
factor(class)21     0.10057      0.16661     0.604   0.54639
factor(class)22     0.09145      0.16082     0.569   0.56988
condLeaftet              NA           NA        NA        NA
condL+quiz               NA           NA        NA        NA
condL+plan               NA           NA        NA        NA
---
Signif. codes:  0 '***' 0.001 '**' 0.01 '*' 0.05 '.' 0.1 ' ' 1

Residual standard error: 0.5393 on 481 degrees of freedom
Multiple R-Squared: 0.1219,     Adjusted R-squared: 0.08353
F-statistic: 3.179 on 21 and 481 DF,  p-value: 3.431e-06
```

The fourth and final approach is the multilevel approach, and it can be done with the **lme** function which is part of the **nlme** package. **lme** stands for linear mixed effect and **nlme** stands for non-linear mixed effect. The **lme** function works like the **lm** function, except that you need to tell it what part of the model is random and what the cluster name is. Here, **random = ~1** means make the intercept random. The **|class** tells the computer that the children are nested within classrooms. We have set **method="ML"**. **ML** stands for maximum likelihood. The default is **REML**, which stands for restricted maximum likelihood. There are disagreements about which of these methods is preferred, but the **ML** method has the advantage that the change in $log($**likelihood**$)$ between models can be compared in a similar manner to the sum of squares in standard ANOVA models. For this reason we keep with **method="ML"** throughout this section.

Here, **model4** is the baseline model and **model4a** includes the effect of condition. They are compared, and we see that the difference is non-significant, $\chi^2(3) = 5.50$, $p = .14$. This is with three degrees of freedom.

```
model4 <- lme(sqw2 ~ 1, random = ~1| class,method="ML")
model4a <- lme(sqw2 ~ cond, random = ~1| class,method="ML")
anova(model4,model4a)
```

	Model	df	AIC	BIC	logLik	Test	L.Ratio	p-value
model4	1	3	836.3542	849.0159	-415.1771			
model4a	2	6	836.8585	862.1820	-412.4292	1 vs 2	5.495703	0.1389

Despite the difference between the models being non-significant, it is still worth examining the coefficient estimates to see if there are any patterns.

```
summary(model4a)

Linear mixed-effects model fit by maximum likelihood
 Data: NULL
       AIC      BIC     logLik
  836.8585 862.182 -412.4292

Random effects:
 Formula: ~1 | class
         (Intercept)  Residual
StdDev:   0.1310681 0.5392071

Fixed effects: sqw2 ~ cond
               Value   Std.Error  DF   t-value p-value
(Intercept) 1.6477056 0.07195975 481 22.897600  0.0000
condLeaftet 0.1881261 0.10466171  18  1.797468  0.0891
condL+quiz  0.1352641 0.10157097  18  1.331720  0.1996
condL+plan  0.2497214 0.10561417  18  2.364469  0.0295
 Correlation:
            (Intr) cndLft cndL+q
condLeaftet -0.688
condL+quiz  -0.708  0.487
condL+plan  -0.681  0.468  0.483

Standardized Within-Group Residuals:
       Min          Q1         Med          Q3         Max
-2.55928601 -0.53230232  0.06347927  0.63347563  2.78461148

Number of Observations: 503
Number of Groups: 22
```

Because **cond** is a **factor**, R will have used the default **contrasts**. These compare the first category with each of the other categories. Therefore, the mean for the control group is 1.65 and each of the three coefficients shows the difference between the control group and each of the experimental conditions. In Chapter 3 there was discussion about **contrasts**. Hill et al. (2007) had specific *a priori* contrasts in which they were interested. These were whether the control group differs from all the leaflet groups; whether the leaflet only group differs from the other two; and whether the leaflet + quiz group differs from the leaflet + plan condition. R has several built-in contrasts but it is often easier to write these in yourself. The following tells R to use the contrasts just described whenever the variable **cond** is used.

```
newcontrasts <- c(-3, 1, 1, 1, 0, -2, 1, 1, 0, 0, -1, 1)
dim(newcontrasts) <- c(4,3)
contrasts(cond) <- newcontrasts
contrasts(cond)
```

```
           [,1] [,2] [,3]
Control     -3    0    0
Leaflet      1   -2    0
L+quiz       1    1   -1
L+plan       1    1    1
```

If the model is re-run with these contrasts, the statistics for the overall fit of the model like AIC and BIC are exactly the same, but the individual coefficients measure different contrasts. We can now see that the control group differs from the leaflet conditions, $t(18) = 2.27, p = .04$. The df for the t test is 18 because there are 22 classrooms and 4 coefficients estimated.

```
model4b <- lme(sqw2 ~ cond, random = ~1| class,method="ML")
summary(model4b)

Linear mixed-effects model fit by maximum likelihood
 Data: NULL
        AIC      BIC     logLik
  836.8585 862.182 -412.4292

Random effects:
 Formula: ~1 | class
          (Intercept)   Residual
StdDev:    0.1310681 0.5392071

Fixed effects: sqw2 ~ cond
                 Value   Std.Error  DF  t-value p-value
(Intercept) 1.7909835 0.03713882 481 48.22403  0.0000
cond1       0.0477593 0.02099840  18  2.27443  0.0354
cond2       0.0014555 0.03083021  18  0.04721  0.9629
cond3       0.0572286 0.05271290  18  1.08567  0.2920
 Correlation:
        (Intr) cond1  cond2
cond1  0.036
cond2 -0.016 -0.009
cond3  0.053  0.032  0.043


Standardized Within-Group Residuals:
        Min           Q1          Med           Q3          Max
-2.55928601  -0.53230232   0.06347927   0.63347563   2.78461148


Number of Observations: 503
Number of Groups: 22
```

Recall from Chapter 4 (on ANCOVA) the discussion of using change scores versus a time 1 score as a covariate. Here, we can increase the power of the comparison by using the time 1 exercise score (also transformed by the square root of exercise plus .5) in the model, and then add condition. This is a multilevel ANCOVA. The result is statistically significant, $\chi^2(3) = 16.67, p < .001$. As mentioned in Chapter 4 you should always look at the interaction between the covariate and any factor. **model5b** does this, and we see adding the interaction does not increase the fit significantly, $\chi^2(3) = 7.33, p = .06$, but the AIC goes down (but the BIC goes up). Given the discussion in Chapter 7 you might consider whether the relationship between the exercise variables are linear. The **bs** function can also be used (for example, **bs(sqw1,df=4)**), but the model does not significantly improve. For more complex multilevel GAMs, the **gamm** function in Wood's (2006, section 6.7) **mgcv** package can be used.

```
model5 <- lme(sqw2 ~ sqw1, random = ~1|class,method="ML")
model5a <- lme(sqw2 ~ sqw1 + cond, random = ~1|class,method="ML")
model5b <- lme(sqw2 ~ sqw1*cond, random = ~1|class,method="ML")
anova(model5,model5a,model5b)
```

	Model	df	AIC	BIC	logLik	Test	L.Ratio	p-value
model5	1	4	400.3260	417.2084	-196.1630			
model5a	2	7	389.6581	419.2022	-187.8290	1 vs 2	16.667947	0.0008
model5b	3	10	388.3317	430.5376	-184.1658	2 vs 3	7.326425	0.0622

Figure 8.2 shows a scatterplot of the two transformed exercise variables with each other. We used the **jitter** function because otherwise there would have been several dots on the same coordinates, and it would not have been possible to tell how many were at each coordinate. The **jitter** function adds a small random error (i.e., a jitter) to each value so that the dots are not on top of one another. We have also let the

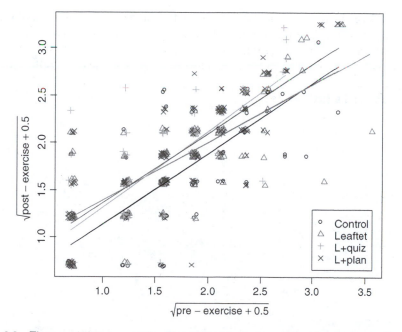

Figure 8.2 The result from a multilevel model of exercise at time 2 using exercise at time 1 as a covariate. This is based on **model5b** where there is an interaction between condition and the covariate

color and symbol type be determined by the condition number (we used the original condition variable, **wcond**, because it is **numeric** while **cond** is a **factor**). We have added **lines** for the predicted values for each condition. The **par** function at the start slightly changes the margins so that the top part of the square root symbol can be seen on the *y* axis label and the **par** function at the end returns them to their default.

```
par(mar=c(5,5,4,2))
plot(jitter(sqw1),jitter(sqw2), xlab=expression(sqrt
  (pre-exercise + .5)),ylab=expression(sqrt(post-exercise + .5)),
  pch=wcond,col=wcond)
legend(3,1.3,c(levels(cond)),pch=1:4,col=1:4)
sexer1 <- split(sqw1,cond)
spred <- split(model5b$fitted[,1],cond)
for (i in 1:4) lines(sexer1[[i]],spred[[i]],col=i)
par(mar=c(5,4,4,2))
```

If you thought making that graph was a lot of work, a simpler graph, that is useful in multilevel modeling, can be made with the **lattice** package. This makes trellis graphs which are very popular among statisticians. Figure 8.3 shows the default **xyplot**; a lot more can be added to it to make it more useful. We only touch upon the graphic capabilities of R, see Murrell (2006) for more details.

```
library(lattice)
xyplot(sqw2~sqw1 | class)
```

The confidence intervals of the different estimates can be found with the **intervals** function. Let's return to the model without using the initial exercise as a covariate. As shown below, the interval for the first contrast, between the control conditions and the others, does not overlap with zero and therefore it is statistically significant.

```
intervals(model5a)
```

Approximate 95% confidence intervals

Fixed effects:
```
                     lower          est.        upper
(Intercept)    0.49439856  0.588418074  0.68243759
sqw1           0.66359295  0.715412254  0.76723156
cond1          0.02713989  0.048667999  0.07019611
cond2         -0.01296502  0.018097991  0.04916100
cond3         -0.04820522  0.005679484  0.05956419
attr(,"label")
[1] "Fixed effects:"
```

  Random Effects:
   Level: class
```
                        lower        est.       upper
sd((Intercept)) 0.01269164  0.04170216  0.1370249
```

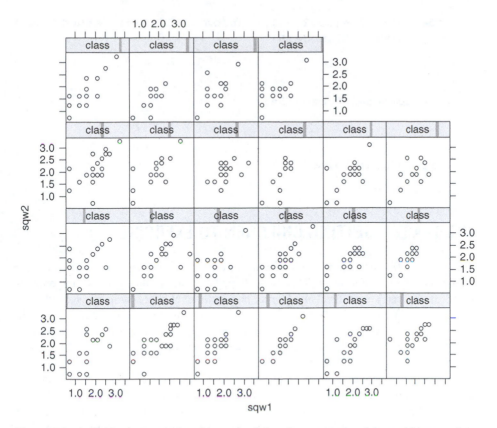

Figure 8.3   Individual scatterplots for each of the classes comparing post-intervention exercise with pre-intervention exercise

```
Within-group standard error:
    lower        est.       upper
0.3279793   0.3493499   0.3721130
```

There is much discussion about an equivalent to R^2 in multilevel modeling, and not much agreement. Several researchers have suggested the proportion of variance accounted for, mostly because of its simplicity. The proportion has to be in comparison with something, and it is not always clear what that something should be. Here is a function that does this if **m1** is the baseline model and **m2** is the model you are testing it against. Various comparisons are made, and the output shows the proportion of variance increasing. It is worth stressing, though, that there are different opinions about the best analog for R^2 in the multilevel case.

```
varprop <- function(m1,m2)
  {v1 <- sum(as.numeric(VarCorr(m1)[,1]))
   v2 <- sum(as.numeric(VarCorr(m2)[,1]))
   rsq <- (v1-v2)/v1
   return(rsq)
  }
```

```
model50 <- lme(sqw2 ~ 1, random = ~1|class,method="ML")
varprop(model50,model5)
```

```
[1] 0.5842596
```

```
varprop(model50,model5a)
```

```
[1] 0.6088032
```

```
varprop(model50,model5b)
```
```
[1] 0.6150124
```

SUMMARY – GETTING CHILDREN TO EXERCISE

Hill et al.'s (2007) data provide a fairly typical example of multilevel data where children are nested within classrooms. This is the type of example that was the impetus for many of the early multilevel modelers. The multilevel regression can be extended to the ANOVA and ANCOVA models, as shown with these data, but also to the other regression extensions described in this book. For example, Hill et al. conducted multilevel mediation analysis (and our comparison of **model5a** and **model5b** above is a multilevel moderator analysis).

The next example has data where the individual is the cluster, and multiple trials are at the lower level. This has become one of the most common examples of multilevel modeling because it can be used in longitudinal methods (Singer & Willett, 2003). In fact, the **nlme** package assumes this is the case (as does the language used in the SPSS multilevel procedures). Within psychology the idea of treating repeated measures data as a multilevel model, and as a more flexible alternative to the random effects ANOVAs, is discussed in Wright (1998) and Wright and London (in press). The next example shows how generalized linear multilevel models can be used instead of methods from signal detection theory (Wright et al., in press).

Example 13 – Response times and accuracy in memory recognition

- Data: Wright et al. (2003).
- Package: **lme4** (Bates & Sarkar, 2007), **e1071** (Dimitriadou et al., 2008).
- Research question: What is the relationship between response time and accuracy for own and other race face recognition?
- Purpose: To show generalized linear multilevel modeling as method for analyzing memory recognition data.

The typical memory recognition study involves showing a set of stimuli and then at a later point asking participants whether they recognize several objects as previously shown. This is usually done by testing people with the originally shown items plus a set of filler items not previously seen. This is called an old/new memory recognition procedure and ten years ago the norm would have been to use signal detection theory (SDT) to differentiate participants' ability to discriminate old from new faces (essentially, accuracy) and a bias to say 'old' (Banks, 1970). Because the standard SDT approach is a form of generalized linear model for each individual (DeCarlo, 1998), it seems natural to analyze these data with a multilevel generalized linear model with individual trial nested within participant. Generalized linear multilevel models are becoming more common in the psychology literature (see Hoffman & Rovine, 2007).

> When analyzing memory recognition data it is tempting to model a response as being correct or incorrect. However, in practice it is often better to model the actual response ('old' or 'new') and use the parameter that denotes whether the object is old or not to estimate accuracy. To see if a variable is associated with accuracy you should test if the interaction between this variable and whether the object is old improves the model. Here, for example, it is expected that the White sample will be more accurate with White faces, and therefore the prediction is an interaction between the race of the face and whether it was previously seen. Sometimes it is convenient to report accuracy, and this is shown done in Figure 8.5.

These data are from the White English participants in Wright et al. (2003). The data are accessed from the book's web page.

```
webreg <- "http://www.sagepub.co.uk//wrightandlondon//"
memrec <- read.table(paste(webreg,"memrec.dat",sep=""),
  header=T)
attach(memrec)
```

Response time measures are usually positively skewed, so this was checked (Figure 8.4). We have used the **paste** function in the code below so that the **skewness** values could be plotted directly onto the graphs. Of course we could have just written the number, but this method allows us to re-use the code for other problems (or if we changed one data point) and it avoids transcription errors.

```
par(mfrow=c(2,1))
hist(time,xlab="Response time (in msec)",
  main="Untransformed variable")
library(e1071)
text(6000,850,paste("skewness =",
  format(skewness(time),digits=2)),pos=4)
lntime <- log(time)
hist(lntime,xlab="ln(response time in msec)",
  main="Transformed variable")
text(8.5,400,paste("skewness =",
  format(skewness(lntime),digits=2)),pos=4)
par(mfrow=c(1,1))
```

The function **lmer** works in a similar way to the **lme** function, but can also be used for generalized linear multilevel models. There are two main differences. First, random effects are shown within the formula so **(1|partno)** tells R that the intercept is random and that the level 2 indicator is the variable **partno**. Thus, **lmer(sqw2~1+(1|class),method="ML")** will estimate the same as **model4** from the last example. This allows models with multiple random variables and non-hierarchical models to be estimated. The second difference is that you are allowed to state the family as with the **glm** command. If it is not stated then normal error with the

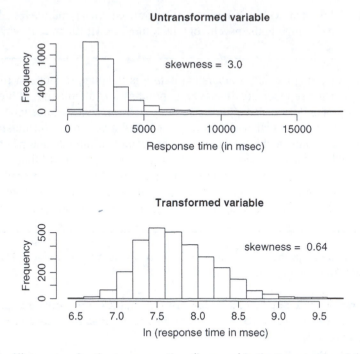

Figure 8.4   Histograms for the response time (in msec) in the top panel and the natural logarithm of response time in the bottom panel. No standard errors or confidence intervals are calculated for skewness because the standard approaches are not appropriate for multilevel data

identity link function is assumed. Here **binomial** is used so the computer assumes it is a logistic regression with binomial error. As with the **glm** command, there are several options for this.

With the linear multilevel models you have the choice between **REML** and **ML**, and we have tended to use **ML** (following Pinhiero & Bates, 2000). For the generalized form, while maximum likelihood is used it has to be approximated, and there are three methods listed in the **lme4** documentation that can do this: **PQL** (for penalized quasi-likelihood), **Laplace**, and **AGQ** (for adaptive Gaussian quadrature). Bates and Sarkar (2007) say that **AGQ** is the most accurate but the slowest and at the time of writing it was not available in this package. **Laplace** is the next most accurate, and following Bates' recommendation is used here.[2]

This first model uses just whether the face has previously been seen to predict participants' responses. It should be large and significant because we would hope that participants are performing above chance.

```
install.packages("lme4")
library(lme4)
```

---

[2]As this book went into print, Laplace became the only method for this procedure. The removed the **method=** option, so if you run the code as is you get a warning. We have not altered the text here because the option to do adaptive Guassian quadrature is likely to be included in future implementations of **lme4**.

```
model1 <- lmer(saysold ~ faceold + (1|partno),
    family=binomial,method="Laplace")
```

The **update** function is used below. The **.** before the **~** means to keep the response variable the same. The **.** after the **~** means keep the model the same but you can add (with a **+**) or remove (with a **-**) any variables. The **var1:var2** means the interaction of these. You can also update other aspects of the model. The following is a sequence of models which add, one at a time: the race of the face, the interaction between the race of the face and whether it is old, a transformed response time, the time by old interaction, the time by race of face interaction, and finally the three-way interaction. The significance of each step can be evaluated with a single **anova** command.

```
model2 <- update(model1, .~. + facewhite)
model3 <- update(model2, .~. + facewhite:faceold)
model4 <- update(model3, .~. + lntime)
model5 <- update(model4, .~. + lntime:faceold)
model6 <- update(model5, .~. + lntime:facewhite)
model7 <- update(model6, .~. + lntime:faceold:facewhite)
anova(model1,model2,model3,model4,model5,model6,model7)
```

```
Data:
Models:
model1: saysold ~ faceold + (1 | partno)
model2: saysold ~ faceold + (1 | partno) + facewhite
model3: saysold ~ faceold + (1 | partno) + facewhite +
   faceold:facewhite
model4: saysold ~ faceold + (1 | partno) + facewhite + lntime +
   faceold:facewhite
model5: saysold ~ faceold + (1 | partno) + facewhite + lntime +
   faceold:facewhite +
model6:      faceold:lntime
model7: saysold ~ faceold + (1 | partno) + facewhite + lntime +
   faceold:facewhite +
model1:      faceold:lntime + facewhite:lntime
model2: saysold ~ faceold + (1 | partno) + facewhite + lntime +
   faceold:facewhite +
model3:      faceold:lntime + facewhite:lntime +
   faceold:facewhite:lntime
```

The spacing on these gets kind of messed up, but it is fairly easy to figure out what the models are.[3] The last few models should have been:

```
model5: saysold ~ faceold + (1 | partno) + facewhite + lntime +
   faceold:facewhite + faceold:lntime
```

---

[3]If you have been playing around with other aspects of R, you may be interested that the command **options(width=150)** does not help, but this may work for future implementations.

```
model6: saysold ~ faceold + (1 | partno) + facewhite + lntime +
    faceold:facewhite + faceold:lntime + facewhite:lntime
model7: saysold ~ faceold + (1 | partno) + facewhite + lntime +
    faceold:facewhite + faceold:lntime + facewhite:lntime +
    faceold:facewhite:lntime
```

```
         Df     AIC      BIC   logLik   Chisq Chi Df Pr(>Chisq)
model1   3   3490.8   3508.8  -1742.4
model2   4   3455.4   3479.5  -1723.7 37.3624      1  9.810e-10 ***
model3   5   3422.2   3452.3  -1706.1 35.2254      1  2.937e-09 ***
model4   6   3424.0   3460.1  -1706.0  0.1894      1     0.6634
model5   7   3384.7   3426.7  -1685.3 41.3408      1  1.279e-10 ***
model6   8   3385.2   3433.3  -1684.6  1.4765      1     0.2243
model7   9   3385.3   3439.4  -1683.7  1.9104      1     0.1669
---
Signif. codes:  0 '***' 0.001 '**' 0.01 '*' 0.05 '.' 0.1 ' ' 1
```

There are lots of different ways to decide on which model looks best. Usually three-way interactions are difficult to explain, and given that the BIC is higher for **model7** than **model6**, and **model6** is higher than **model5**, it seems best to treat **model5** as the most useful.

This model is examined by typing its name:

model5

```
Generalized linear mixed model fit using Laplace
Formula: saysold ~ faceold + (1 | partno) + facewhite + lntime +
faceold:facewhite + faceold:lntime
 Family: binomial(logit link)
  AIC   BIC  logLik deviance
3385  3427   -1685     3371


Random effects:
 Groups Name        Variance Std.Dev.
 partno (Intercept) 0.11544  0.33977

number of obs: 3000, groups: partno, 50

Estimated scale (compare to  1 )  0.9899662

Fixed effects:
                  Estimate Std. Error z value Pr(>|z|)
(Intercept)        -5.8387     1.1447  -5.100 3.39e-07 ***
faceold            10.7439     1.4355   7.485 7.18e-14 ***
facewhite          -1.0359     0.1331  -7.780 7.24e-15 ***
lntime              0.6486     0.1454   4.460 8.18e-06 ***
faceold:facewhite   0.9677     0.1742   5.555 2.78e-08 ***
```

```
faceold:lntime      -1.1774      0.1830   -6.436 1.23e-10 ***
---
Signif. codes:   0 '***' 0.001 '**' 0.01 '*' 0.05 '.' 0.1 ' ' 1

Correlation of Fixed Effects:
             (Intr) faceld facwht lntime fcld:f
faceold      -0.741
facewhite    -0.096  0.071
lntime       -0.997  0.739  0.055
facld:fcwht   0.073 -0.082 -0.764 -0.042
faceld:lntm   0.737 -0.997 -0.038 -0.739  0.031
```

Under Random effects, the variance and standard deviation associated with **partno** is listed (.115 and .340, respectively). This is the variability around the intercept. So, assuming normality, about 95% of the population should be between about −6.5 and −5.0 (i.e., −5.8 ± 2 × .34).

In general psychologists usually focus on the fixed effects, and begin with the interactions. **faceold:facewhite** means that there is an own race bias. These White participants were more accurate (meaning whether the item was seen was more predictive of whether they said seen) with White faces than with Black faces. The significance of the **faceold:lntime** effect means that response time also predicted accuracy. That it is negative means longer times were associated with more errors.

Figure 8.5 was made to look at the relationship between response time and accuracy. First the predicted probabilities from the estimates of the fixed effects from above

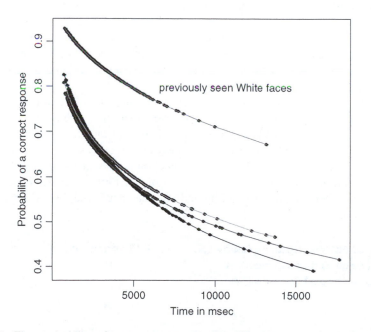

Figure 8.5   The probability of a correct response for different response times, based on model 5. Controlling for response time, previously seen White faces are those most accurately recognized

are calculated. The predicted probability is made by taking any value x and finding: $e^x/(1 + e^x)$.

```
mod5 <-   -5.8387 + faceold*10.7439+facewhite*-1.0359
   +lntime*0.6486+faceold*facewhite*0.9677+faceold*lntime*-1.1774
predprob <- exp(mod5)/(1+exp(mod5))
```

We decided to show the probability of a correct response rather than an old response, so a variable **rightprob** is created. **faceold** is 0 if new and 1 if old, so this flips the probabilities for the new faces around.

```
rightprob <- predprob*faceold + (1-faceold)*(1-predprob)
```

The following code makes the graph. For the **lines** command it was necessary to **sort** the data for **stime** and to place the values for **srightprob** in this **order**.

```
par(mfrow=c(1,1))
plot(time,rightprob,pch=20,xlab="Time in msec",ylab="Probability
   of a correct response",cex.lab=1.3)
stime <- split(time,facewhite+2*faceold)
srightprob <- split(rightprob,facewhite+2*faceold)
for (i in 1:4)
lines(sort(stime[[i]]),srightprob[[i]][order(stime[[i]])],col=i,
   lwd=1.5)
text(12000,.8,"previously unseen White faces",cex=1.3)
```

Figure 8.5 is a fairly interesting graph. It shows for all four conditions (new and old faces, White and Black faces) that as response time increased the probability of a correct response decreased. Also, the line for White old faces stands out. Controlling for response time, these are the most accurate.

The next model allows the level of accuracy (the coefficient for the **faceold** variable) to vary by participant. The **update** function is used. First you remove the random intercept (**-(1|partno)**) and then you add the random variable for **faceold** (**+ (faceold|partno)**), which also includes the intercept (this can be done in other ways, too). The **anova** function shows that this model fits better, $\chi^2(2) = 23.77, p < .001$. The 2 degrees of freedom are for the variance in accuracy and the correlation between this variability and the intercept variability. The fixed effects remain approximately the same.

```
model5b <- update(model5, .~. -(1|partno) +
   (faceold|partno))
anova(model5,model5b)
```

```
Data:
Models:
model5: saysold ~ faceold + (1 | partno) + facewhite + lntime +
   faceold:facewhite +
model5b:      faceold:lntime
```

```
model5:  saysold ~ faceold + facewhite + lntime + (faceold | partno) +
model5b:       faceold:facewhite + faceold:lntime
```

```
        Df    AIC     BIC   logLik  Chisq Chi Df Pr(>Chisq)
model5   7  3384.7  3426.7 -1685.3
model5b  9  3364.9  3419.0 -1673.5 23.768      2  6.901e-06 ***
---
Signif. codes:  0 '***' 0.001 '**' 0.01 '*' 0.05 '.' 0.1 ' ' 1
```

model5b

```
Generalized linear mixed model fit using Laplace
Formula: saysold ~ faceold + facewhite + lntime +
   (faceold | partno) +
faceold:facewhite + faceold:lntime
 Family: binomial(logit link)
  AIC  BIC logLik deviance
 3365 3419  -1673     3347
Random effects:
 Groups Name         Variance Std.Dev. Corr
 partno (Intercept) 0.35606  0.5967
        faceold     0.49999  0.7071   -0.839
number of obs: 3000, groups: partno, 50
```

```
Estimated scale (compare to  1 )  0.9790407
```

```
Fixed effects:
                   Estimate Std. Error z value Pr(>|z|)
(Intercept)        -7.5122     1.2431   -6.043 1.51e-09 ***
faceold            13.3075     1.5900    8.370  < 2e-16 ***
facewhite          -1.0640     0.1355   -7.850 4.16e-15 ***
lntime              0.8581     0.1578    5.439 5.37e-08 ***
faceold:facewhite   0.9930     0.1761    5.639 1.71e-08 ***
faceold:lntime     -1.5012     0.2024   -7.416 1.21e-13 ***
---
Signif. codes:  0 '***' 0.001 '**' 0.01 '*' 0.05 '.' 0.1 ' ' 1
```

```
Correlation of Fixed Effects:
            (Intr) faceld facwht lntime fcld:f
faceold     -0.781
facewhite   -0.088  0.069
lntime      -0.996  0.778  0.050
facld:fcwht  0.068 -0.077 -0.770 -0.038
faceld:lntm  0.775 -0.995 -0.039 -0.779  0.032
```

   The shape of the curves in Figure 8.5 is dependent on the log transformation used, the response time variable, and the logistic model used in the regression. Without these the actual model is a straight line. It is unlikely that the relationship is this simple.

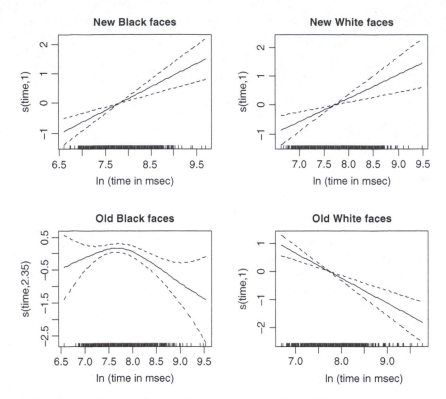

Figure 8.6 Generalized additive multilevel models for the relationship between response time and the probability of responding 'old'. The assumption in previous models was that these would be linear (after accounting for the link function and the initial transformation of the data), which appears true for three of the four conditions

There is a generalized additive multilevel model (**gamm**) function written for R. It is available in Simon Wood's (2006) **mgcv** package. This could be used to investigate this relationship in more detail. It requires the **s** function for the spline rather than the **bs** function used previously. The default for **s** is a cubic spline with one knot. The following code makes Figure 8.6. The **gamm** function would not allow us to use **ssaysold[[i]]** within it, which is why it is done before the **gamm** function. As Figure 8.6 shows, the linear relationship assumed for Figure 8.5 looks valid for three of the four conditions.

```
install.packages("mgcv")
library(mgcv)
par(mfrow=c(2,2))
ssaysold <- split(saysold,facewhite+2*faceold)
slntime <- split(lntime,facewhite+2*faceold)
spartno <- split(partno,facewhite+2*faceold)
for (i in 1:4) {
    part <- spartno[[i]]
    time <- slntime[[i]]
    sold <- ssaysold[[i]]
```

```
gammy <- gamm(sold ~ s(time),random=list(part=~1),
    family=binomial)
plot(gammy$gam,xlab="ln(time in msec)")
if (i == 1) title("New Black faces")
if (i == 2) title("New White faces")
if (i == 3) title("Old Black faces")
if (i == 4) title("Old White faces")
}
```

SUMMARY – MEMORY RECOGNITION

While methods from signal detection theory (SDT) are common for memory recognition studies, they present two difficulties (Wright et al, in press). The first is that they are usually done by first calculating measures (like d') for each individual and then using these aggregate measures in analysis. This is the aggregating approach from the Hill et al. example and can fall foul of the ecological fallacy, it lowers the effective sample size, it ignores differences in weighting between people, etc. Second, if you are interested in a covariate, like response time, the standard SDT methods are difficult to implement since you have to split that continuous covariate into bins and create SDT measures for each person for each bin. This can create lots of problems. However, the multilevel modeling approach is well suited for this. Because the standard SDT model is a generalized linear model (DeCarlo, 1998), it is a simple extension to treat participant as a random variable and run a multilevel generalized linear model. While SDT has a long history within memory research, and has been very useful in theory construction, it is likely that the flexibility of approach used here will mean multilevel models become the norm for recognition data in years to come. In addition, further extensions to this model can be made. For example, you can allow a random effect for the face using (1|face) which usually improves the model (see Wright & London, in press, for more details).

SUMMARY

Multilevel modeling is one of the hot methods now in lots of areas of science, including psychology. While the traditional example has been with people nested within larger clusters (so, pupils nested within classrooms), because of the great amount of medical research with multiple measurements per person, multilevel models with the person as the higher order level are now common (perhaps more common). Harvey Goldstein, one of the pioneers of this approach, talks about how there are hierarchies everywhere you look. Multilevel modeling is now one of the tools expected for social and psychological scientists.

We will end with a caveat. While multilevel models are now expected to be used in areas where the hierarchical structure is obvious, more research is necessary to see how useful they are when the levels are not such clean structures and where the components at different levels cannot be viewed as some random sample of those at that level. This was essentially Jacob Cohen's (1976) criticism of Herb Clark's (1973) language as a fixed effect fallacy. Perhaps some of the resampling techniques (and local causal inference) will be applied to these situations. As with all statistical procedures, it is critical to examine the data carefully and consider the alternatives before running any statistical test.

SOME WORDS/CONCEPTS WORTH REMEMBERING

R concepts

- more on doing contrasts within R;
- estimation methods for multilevel linear and multilevel generalized linear models.

R functions

- `aggregate`: to calculate group measures (see also `tapply`);
- `lme`: linear mixed effect models;
- `nlme`: non-linear mixed effect models;
- `par(mar=`: for changing a graph's margin;
- `lattice`: a graphics package within R;
- `xyplot`: for scatterplots of different groups;
- `intervals`: prints the confidence intervals of `lme`/`nlme` models;
- `lmer`: generalized linear mixed effect models;
- `gamm`: generalized additive mixed effect models.

Statistical concepts

- multilevel models: models for clustered data;
- SDT: Signal detection theory.

FURTHER READING

http://www.cmm.bristol.ac.uk/ is the multilevel modelling centre's web page and has a wealth of information on the topic.

Goldstein, H. (2003). *Multilevel statistical methods (3rd edition)*. London: Edward Arnold. This is the bible of multilevel modeling, but is a bit heavy on the statistics.

Hox, J. (2002). *Multilevel analysis: Techniques and applications*. London: Lawrence Erlbaum Associates. This book covers lots of topics and would be a good book for a one-term graduate course for psychologists.

Kreft, I. I. & de Leeuw, J. (1998). *Introducing multilevel modeling*. London: Sage Publications. This is one of the clearer introductions to multilevel modeling, focusing on linear models. This is an excellent book and is more introductory than any of the others.

Singer, J. D. & Willett, J. B. (2003). *Applied longitudinal data analysis: Modeling change and event occurrence*. New York: Oxford University Press. This is a book that covers longitudinal methods. The first half focuses on multilevel models where the individual testing session is nested within the person. The writing is really clear. They have a book on multilevel modeling currently in preparation.

Sullivan, L. M., Dukes, K. A. & Losina, E. (1999). An introduction to hierarchical linear modelling. *Statistics in Medicine, 18*, 855–888. This review is aimed more towards medical researchers.

Wright, D. B. (1998). Modelling clustered data in autobiographical memory research: The multilevel approach. *Applied Cognitive Psychology*, *12*, 339–357. This is one of the first introductions aimed at psychologists. It covers both linear and generalized linear multilevel models.

Wright, D. B. & London, K. (in press). Multilevel modelling: Beyond the basic applications. *British Journal of Mathematical and Statistical Psychology*.

9

Robust regression

Learning outcomes

1. That the standard approaches to statistical inference are highly influenced by outliers and lack power under most empirical conditions.
2. That there are several alternative procedures including:

   - rank based procedures;
   - eliminating outliers;
   - M-estimates.

In 1805 Adrien Marie Legendre introduced the idea of minimizing the square of the residuals: 'it consists of making the sum of the squares of the errors a *minimum* … it prevents the extremes from dominating' (translation from Stigler, 1986: 13, original French manuscript printed on p. 58).[1] The ease of computing least squares, its conceptual appeal, and the fact that least squares estimation is well suited for a very particular (and rare) set of situations means that this approach has dominated statistics. Every time we calculate a mean, *t* test, ANOVA, etc., we are minimizing the sum of the squared residuals. The least squares approach is one of several possible *loss functions*.

The second part of the quote from Legendre deserves further scrutiny. Squaring a residual means the impact of large residuals will be greater than if, for example, the absolute value was taken (an approach that actually pre-dated Legendre, but is computationally difficult so was not widely used until recently). Small residuals have little impact on least squares, but as their value increases the impact becomes very large. Large residuals are not as influential for minimizing the sum of absolute values. Another method is to trim data beyond a certain value. For example, the 20% trimmed mean is the

---

[1]Legendre and 1805 are generally given as the person and date for the introduction of least squares, although Gauss probably was using it since 1795. Soon after 1805, Gauss did publish a much extended formulation of least squares. Stigler (1999: 331) concluded that while Gauss may have discovered least squares, it was Legendre 'who first put the method within the reach of the common man.'

mean of values excluding the extreme 20% from each end of the scale.[2] The R function **mean** allows trimming as an option, so **mean(xvar,.2)** produces the 20% trimmed mean (and **mean(xvar,.5** produces the same value as **median**). All three of these loss functions (least squares, least absolute values, and trimming) can be applied to estimating any quantitative parameter.

The second part of Legendre's quotation is wrong; extremes can dominate with least squares estimation. Robust alternatives lessen the impact of these extremes. Consider the following three datasets:

1. Set 1: 1, 2, 3, 4, 5, 6, 7, 8, 9, 10

    - Mean $= 5.5$, 95% CI $= (3.33, 7.67)$,
    - Standard deviation $= 3.03$,
    - $t(9) = 5.75$, $p < .001$.

2. Set 2: 1, 2, 3, 4, 5, 6, 7, 8, 9, 100

    - Mean $= 14.5$, 95% CI $= (-7.07, 36.07)$,
    - Standard deviation $= 9.54$,
    - $t(9) = 1.52$, $p = .16$.

3. Set 3: 1, 2, 3, 4, 5, 6, 7, 8, 9, 1000

    - Mean $= 104.5$, 95% CI $= (-120.59, 329.60)$,
    - Standard deviation $= 314.66$,
    - $t(9) = 1.05$, $p = .32$.

As the most extreme point gets larger and larger, the mean goes from 5.5 to 14.5 to 104.5. This compares with minimizing the sum of least absolute values (which produces the median for univariate analyses) and the 20% trimmed mean, which both remain at 5.5. The outlier has less impact with these robust estimators. Depending on your research question, you may or may not want a single point to have this much impact.

How does the outlier impact on hypothesis testing and confidence interval estimation? Suppose we wanted to test the hypothesis that the data above come from a population with a mean of 0. A common approach is to calculate a t test: $t = \bar{x}\sqrt{n}/sd$ and the corresponding 95% confidence interval: $\bar{x} \pm t_{crit} \cdot sd/\sqrt{n}$. Because the outlier affects the standard deviation even more than the mean (see above), as the outlier moves away from the null hypothesis it actually makes the statistic *less* significant. It makes the estimate much less precise, as is reflected in the confidence intervals. While this may seem paradoxical, it is not a new discovery (see Fisher, 1925: 112, for a similar example). It is just that only recently have robust procedures become widely available.

A topical example showing the effect of removing an outlier was published the week before the 2004 US presidential election and concerned the number of civilians deaths in Iraq since the younger George Bush's war there (Roberts et al., 2004; a more recent survey is Burnham et al. (2006) where they do not discuss the problem of this outlier). The most discussed statistic they report is an estimate that 98,000 more civilians died during the post-invasion occupation than expected, though the 95% confidence interval

---

[2]If you are calculating the trimmed mean, do not simply exclude the extremes and conduct analyses as if the data were not trimmed. The equations to estimate the standard error (and therefore p values) are different (Wilcox, 2003a).

was large, from 8,000 to 194,000. This was based on a pre-invasion mortality rate of 5.0 (per 1,000 per year) with an interval from 3.7 to 6.3. They estimated the post-invasion mortality rate to be 12.3 with an interval from 1.4 to 23.2, but did not use this estimate to reach their conclusions. As the interval includes the estimated pre-invasion mortality rate, it does not provide strong evidence for an increase. Instead they removed the data of Falluja, where the fighting was most intense. This lowered the estimated mortality rate to 7.9 but made the interval smaller, from 5.6 to 10.2, thereby providing better evidence for an increased mortality rate. There is an important political point raised by this paper, why were the occupying forces not doing more to keep track of the number of civilian deaths? The US General Tommy Franks reportedly said (Roberts et al., 2004: 1863): 'we don't do body counts,' apparently as a snub of Geneva Convention guidelines for an occupying force. The statistical point, that eliminating an outlier can make the confidence intervals much smaller, raises an ethical concern. Given that the researchers knew Falluja was going to be an outlier and therefore that they were likely to exclude it from many of their analyses, and that Falluja was a very (very) dangerous place for their researchers to be operating, should they have been trying to gather data from there?

The Iraqi data and the three data sets above are clearly not normally distributed. Many people believe that: a) psychology data sets are usually normally distributed and b) if they were not, we would notice any discrepancy large enough to matter. Micceri (1989) surveyed a large number of psychology data sets. He found *none* approximated the normal distribution and most were very un-normal. But would you be able to notice if a distribution was un-normal enough to matter? The main part of Figure 9.1 shows two

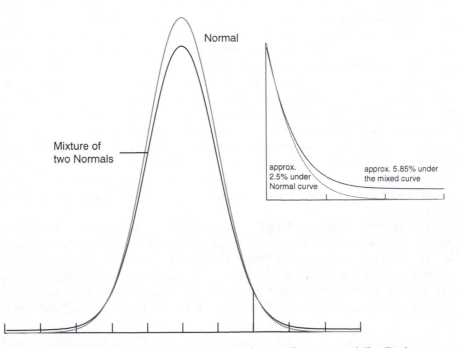

Figure 9.1   A normal distribution in gray and a mixture of two normal distributions (90% with a sd = 1, and 10% with an sd = 10) in black. The tails of these distributions are shown in the upper right hand corner

curves. One is normally distributed; one is not. The mixture curve (which is 90% a normal curve with a standard deviation of 1 and 10% a normal curve with a standard deviation of 10) looks very similar to the standard normal curve. If you observed data that looked like this mixture distribution, you would be likely to assume that the normal distribution assumption had been met. Tukey (1960) showed that these distributions differed in a very important way. The interest, particularly for null hypothesis significance testing (which is dominant in psychology), is often in the tails of the distribution. The upper-right hand corner of Figure 9.1 shows the tails of these distributions. The mixture distribution has a lot more area under the curve beyond $z = 1.96$.

Because of this area under the curve and that the outliers can affect measures of precision more than location, the standard least squares procedure is often less able to identify differences. This led Rand Wilcox to ask the following question as the title of an *American Psychologist* paper: 'How many discoveries have been lost by ignoring modern statistical methods?' (1998). In most situations, least squares procedures are less powerful than methods which are less influenced by outliers, despite what is said in many textbooks. This means that often people using traditional methods will be missing significant effects. Wilcox poignantly makes this point in another provocative title: 'ANOVA: A paradigm for low power and misleading measures of effect size?' (1995).

We have mentioned least absolute values and trimmed estimates. Trimmed statistics are fairly popular. They are conceptually simple and have good properties. Much of the discussion in Wilcox (2003a) is about trimmed estimates. Statisticians have come up with other robust loss functions. The most popular of these are M-estimators (there are also R-, L-, S-, and W-estimators; see Maronna et al., 2006; Wilcox, 2003a, for details). The largest collection of robust procedures has been written for R/S-Plus (http://cran. r-project.org/ and http://lib.stat.cmu.edu/S/). The procedures that come with the package allow robust GLMs (and therefore linear regressions too), ANOVAs, correlations, principal component analysis, etc. Other packages like SPSS and SYSTAT also include some M-estimators in some of their procedures.

Robust procedures increase the likelihood of finding a significant result; they are endorsed by the APA task force on statistics (Wilkinson et al., 1999), and are becoming more common in the main packages psychologists use. They will become more popular.

We cover three robust methods. The first is already used by many psychologists: Spearman's ρ (rho) correlation which is based on ranked data. The example used to illustrate this procedure is about crime in neighborhoods in Sussex (UK). We discuss some problems with this measure, and then go through an example concerning children's well-being in several wealthy nations. We use the skipped correlation, which we like for its conceptual simplicity. It also allows us to direct readers to a very useful collection of functions on Wilcox's web page. The final example uses the best made-up data set ever (Anscombe, 1973) and illustrates M-estimator robust regression.

Example 14 – Crime in Sussex

- Data: Published in *The Argos* (local Brighton newspaper on 10 March, 2007, pp. 6–8).
- Research question: What is the association between drug offences and thefts for the Sussex neighborhoods?
- Purpose: To show how Spearman's ρ is less influenced by univariate outliers than Pearson's r.

The crime statistics in Sussex, England, for 2005–6 were recently published in the local Brighton paper and broken down by neighborhood and type of offence. They are available on the book's web page:

```
webreg <- "http://www.sagepub.co.uk//wrightandlondon//"
crime <-
read.table(paste(webreg,"sussexcrime.dat", sep=""),header=T)
```

We order the data set by theft (in the 9th column) which will be useful for some of the graphs. This command re-orders the entire data set (columns **1:10**) as is done when ordering a data set with many of the main statistical packages.

```
crime <- crime[order(crime[,9]),1:10]
attach(crime)
```

The left hand panel of Figure 9.2 shows for the different neighborhoods the number of thefts and drug offences. Clearly there is a positive relationship between the two types of offences, but one point stands out: 'Regency', which corresponds to the center of Brighton, where people drink, get high, and thieve. Regency is an outlier for both crimes because it has far more of both types. Similar graphs are made for the data when ranked and also when logged.

```
par(mfrow=c(1,3))
plot(theft, Drugs, xlab="Theft offences",ylab="Drug
    offences",pch=19,main="Scatterplot of raw data",
    cex.lab=1.3)
text(2900,300,"Regency",pos=3,cex=1.3)
plot(rank(theft), rank(Drugs) ,xlim=c(0,300),
    ylim=c(0,300), xlab="Rank of theft offences",
    ylab="Rank of drug offences",pch=19,
    main="Scatterplot of ranked data",cex.lab=1.3)
```

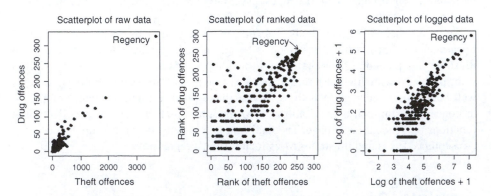

Figure 9.2   Scatterplots of different Sussex neighborhoods by the number of drug offences and thefts in 2005–6. The left panel shows the raw data. The middle panel shows the ranked data and includes a small jitter so that all the points can be seen. The right panel shows the data after using ln transformation on both variables. As is common, we took the ln of the variable plus 1

```
text(180,270,"Regency",pos=3,cex=1.3)
arrows(230,285, 255,265, length=.07)
plot(log(theft+1), log(Drugs+1), xlab="Log of theft
    offences + 1",ylab="Log of drug offences + 1",
    pch=19,main="Scatterplot of logged data",cex.lab=1.3)
text(6.7,5.4,"Regency",pos=3,cex=1.3)
par(mfrow=c(1,1))
```

Spearman's ρ addresses problems with cases like Regency which are univariate outliers. By univariate outlier, we mean that the case is an outlier just looking at the drug variable, and it is an outlier just looking at the theft variable. Spearman's ρ involves taking the ranks of each of the variables on their own, and then conducting Pearson's r on the ranks. Ranking procedures, like this, are a popular method for analyzing data in psychology when you do not believe either that the data are normally distributed or that you do not believe they are interval (Conover & Iman, 1981). Ranking procedures are popular, due in part to Siegel (1956 and later editions) providing clear 'how-to' descriptions of some of these tests. While there have been advances in ranked-based procedures since the 1950s (Cliff & Keats, 2002), those described in Siegel remain the most popular in psychology.

The R function **cor.test** (used also in Chapter 5) calculates both Pearson's and Spearman's correlations:

```
cor.test(theft,Drugs)

Pearson's product-moment correlation

data: theft and Drugs
t = 43.9704, df = 259, p-value < 2.2e-16
alternative hypothesis: true correlation is not equal to 0
95 percent confidence interval:
 0.9228864 0.9519523
sample estimates:
      cor
0.9390764


cor.test(theft,Drugs,method="spearman")

Spearman's rank correlation rho

data: theft and Drugs
S = 656025.4, p-value < 2.2e-16
alternative hypothesis: true rho is not equal to 0
sample estimates:
      rho
0.7786106


Warning message:
Cannot compute exact p-values with ties in: cor.test.
    default(theft, Drugs, method = "spearman")
```

These statistics are: $r = .94$, $p < .001$, $n = 261$, and $\rho = .78$, $p < .001$, $n = 261$. The value for Spearman's is much smaller. Much of this is due to Regency (without this outlier, r drops to .89). Most psychology journals either recommend or require confidence intervals to be reported. R does not print a confidence interval for Spearman's ρ. One possibility is finding a bootstrap estimate of the confidence interval. The **boot** function requires the data to be placed into a single object. When doing bootstrap estimates for the **median** and **skewness** in other chapters only a single variable was being analyzed at a time. Because two variables are required for Spearman's ρ the variables **Drugs** and **theft** are combined into the object **thevars**. The BCa estimate for the interval is between .71 and .84.

```
library(boot)
bootspear <- function(x,i)
cor.test(x$Drugs[i],x$theft[i],
    method="spearman")$estimate
thevars <- as.data.frame(cbind(Drugs,theft))
spearboot <- boot(thevars,bootspear, R=1000)
boot.ci(spearboot)

BOOTSTRAP CONFIDENCE INTERVAL CALCULATIONS
Based on 1000 bootstrap replicates

CALL :
boot.ci(boot.out = spearboot)

Intervals :
Level       Normal                    Basic
95%      ( 0.7185, 0.8427 )   ( 0.7212, 0.8483 )

Level       Percentile                BCa
95%      ( 0.7089, 0.8361 )   ( 0.7078, 0.8356 )
Calculations and Intervals on Original Scale
Warning message:
bootstrap variances needed for studentized intervals in:
boot.ci(spearboot)
```

An alternative is simply to rank the variable and find the confidence interval for Pearson's r on the ranks. Most methodologists would prefer the bootstrap estimates, but all these estimates are similar.

```
cor.test(rank(theft),rank(Drugs))

Pearson's product-moment correlation

data: rank(theft) and rank(Drugs)
t = 19.9688, df = 259, p-value < 2.2e-16
alternative hypothesis: true correlation is not equal to 0
95 percent confidence interval:
 0.7258077 0.8222921
```

```
sample estimates:
      cor
0.7786106
```

Ranking the variables lessens the impact of any univariate outlier. Thus, Pearson's r on raw data is $r = .94$, but it is very influenced by this one data point and also by a couple of the other HTDs (the sociologists' abbreviation of *havens for thieving druggies*). Ranking the data on both of these variables means that all the wholesome areas that are squished into the lower left-hand corner of the first panel of Figure 9.2 are spread out in the second panel, and Regency and other HTDs are pulled in. When the correlation is run on the ranks, what is Spearman's ρ, you get .78. As can be seen in the output above, R gives a warning that because there are ties exact p values are not computed. The final command shows that you get the same correlation when directly calculating Pearson's r on the ranked data.

There are a couple of difficulties with Spearman's ρ. The first is the same as with the other ranked based procedures. Ranking is a particular transformation and when it is done all meaning about the distances between adjacent points is lost. You would know that there were more drug offences in Regency than elsewhere, but you would not know how much more and there is nothing you could do with these ranks to get back to the original data. The inference becomes about the ranks of data, and this can make it difficult to describe the results.

Here, a better alternative to lessen the impact of HTDs might be to take the natural logarithms (*ln*) of the variables plus 1 (the $+1$, which is sometimes called a starting value, prevents negative infinities for the places with 0 of a particular type of offence; Mosteller & Tukey, 1977: 91). For these transformed values, $r = .76$ and $\rho = .74$, and Regency no longer stands out (see right panel of Figure 9.2). The *ln* transformation is a useful transformation for many positively skewed variables (so is the square root transformation though neither of these work with negative values unless a starting value is added to the original variable, see also the Box-Cox family of transformations, 1964). Figure 9.3 shows the line through the scatterplot of the logged data (left panel) and the back-transformed line through the raw data. The regression on the raw data is shown in gray, so it can be seen how it is more influenced by the outlier than the logged regression.

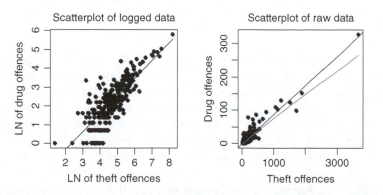

Figure 9.3   A scatterplot of the logged data with the logged regression line (left panel), and a scatterplot of the raw data with the back-transformed predicted values from the logged regression. The regression on the raw data is shown in black

The code for Figure 9.3 is shown below. Note that the transformations are actually done within the **lm** and **lines** functions.

```
lnreg <- lm(log(Drugs+1)~log(theft+1))
par(mfrow=c(1,2))
plot(log(theft+1), log(Drugs+1), xlab="LN of theft
    offences",ylab="LN of drug offences",pch=19,
    main="Scatterplot of logged data",cex.lab=1.3)
lines(log(theft+1),predict(lnreg))
plot(theft, Drugs, xlab="Theft offences",
    ylab="Drug offences",pch=19,main="Scatterplot of
    raw data",cex.lab=1.3)
lines(theft,exp(predict(lnreg))-1)
abline(lm(Drugs~theft),col="black")
par(mfrow=c(1,1))
```

The second problem with Spearman's ρ, and it also exists for the *ln* transformation (and for any transformation that looks only at each variable individually), is that they just examine univariate outliers. Given that the point of regressions is to examine the relationships among variables, it would be good to have a technique that can look for bivariate (and multivariate) outliers and lessen their influence. There are several techniques that can do this, and we opt to present a conceptually simple one called the *skipped correlation coefficient* (Wilcox, 2003b). There are actually a variety of skipped correlations, but we will use the default described by Wilcox. This is a new procedure, and while there are more computationally advanced methods, this one works fairly well, there is an R function for it (**scor**) and it is described in the next example.

Example 15 – Children's well-being

- Data: From United Nations Children's Fund (2007) report.
- Packages: Functions from Rand Wilcox's web page are accessed.
- Research question: What is the relationship between a country's health and risk scores?
- Purpose: To illustrate how the *skipped correlation coefficient* limits the impact of bivariate outliers.

The United Nations Children's Fund's (UNICEF, 2007) report card on children of 21 nations ranked the nations on 7 different attributes, including health and risk. These data are already ranked from 0 to 20, where 0 means good and 20 means bad. So Sweden is the point down in the lower left-hand corner, and US is the point in the top right-hand corner of Figure 9.4. The UK is the one with the worst risk score, preventing the US from being the worst on both – there have been some criticisms of these data.

The data are on the book's web page and can be accessed in the usual way:

```
webreg <- "http://www.sagepub.co.uk//wrightandlondon//"
children <- read.table(paste(webreg,"unicef.txt",
    sep=""),header=T)
attach(children)
```

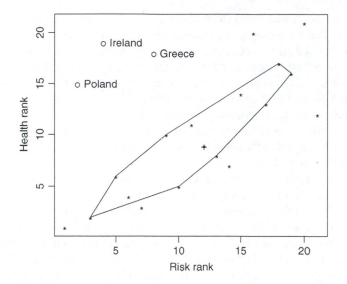

Figure 9.4   The ranks of health behaviors with risk behaviors for 21 wealthy countries (UNICEF, 2007). The polygon (i.e., the enclosed shape) in the middle was made by Wilcox's (2003b) skipped correlation function in R to show where about half of the data are. The + shows the mean for the two variables once the outliers are excluded

Pearson's r and Spearman's ρ can be found as:

```
cor.test(risks,health)

Pearson's product-moment correlation

data: risks and health
t = 2.6049, df = 19, p-value = 0.0174
alternative hypothesis: true correlation is not equal to 0
95 percent confidence interval:
 0.1044252 0.7734038
sample estimates:
      cor
0.512987
```

```
cor.test(risks,health,method="spearman")

Spearman's rank correlation rho

data: risks and health
S = 750, p-value = 0.01857
alternative hypothesis: true rho is not equal to 0
sample estimates:
     rho
0.512987
```

Since the data are already ranked, Pearson's *r* and Spearman's ρ are the same (.51), which means that according to these statistics there is a positive relationship between the two, and that the effect, in Cohen's (1988) terms, is 'large'. This is an example where 'large' is a misnomer. As discussed in Chapter 8, when you have data where each data point is based on lots of people, the effect sizes tend to be much larger than when based on their individual constituents (Robinson, 1950).

The problem with these estimates is that while there are no univariate outliers (because the data are already ranked), there may be bivariate outliers. These are points where the combined values do not fit the pattern of the rest of the data. If the data were the length of the right foot and the left foot of a group of people, finding a right foot that is 10 inches long is not striking, nor is finding a left foot that is 6 inches long, but to find these together on the same person would be surprising. Figure 9.4 shows a scatterplot between risks and health. Three of the countries have been labeled because they stand out. This figure was made with the **scor** function from Wilcox (2003b). The **source** command below accesses his functions from his website. We have no control over this website, so they may move locations. If the command fails to work try googling 'Rand Wilcox' and find where they are. We will update our web pages accordingly. The **scor** function is in the second set of functions accessed, but it calls functions from the first set, so both need to be accessed.

```
source("ftp://ftp.usc.edu/pub/wilcox/Rallfunv1.v4")
source("ftp://ftp.usc.edu/pub/wilcox/Rallfunv2.v4")
scor(risks,health,xlab="Risk rank",ylab="Health rank")

$cor
[1]  0.8573386

$test.stat
[1]  7.259896

$crit.05
[1]  2.650510
```

The **scor** function makes the scatterplot and you can add in other information to the plot. It is often worth labeling outliers, and therefore we have done this with the **text** function.

```
text(c(2,8,4),c(15,18,19),c("Poland","Greece",
    "Ireland"),pos=4)
```

Figure 9.4 shows this positive relationship but it also shows that the data points for Ireland, Greece, and Poland do not fit with this trend. These countries score fairly well on 'risks', but poorly on 'health'. These outliers will have a large impact on the standard regression and correlation, and the question is whether you want them to. The move within statistics is that you should try to lessen their impact, but before automatically following this belief (which we share), you should ask yourself whether these points are particularly important and are there any explanations for why they are different. If in fact they are qualitatively different (like someone sneezing in a reaction time task), then you probably do not want to include them in your sample because they are not part of your population of interest. Presumably most cognitive theories are not trying to account for cognition while sneezing. However, often outliers are important, so they are worth careful examination.

The first step for the skipped correlation is to decide which data points are outliers. In Figure 9.4 it appears that there are three outliers, but it is worth having a general rule with good statistical properties. Wilcox uses a complex algorithm based on where the bulk of the data lie. His **scor** function makes the plot shown in this figure with the outliers shown with circles, a polygon containing the bulk of the data, and a + for the mean of both variables excluding the outliers. It decides the circle points are outliers, removes these, and runs the correlation (you can use Pearson's or Spearman's, here Pearson's was used).[3] You get $r_{skip} = .857$, which is much higher than that found with the standard Pearson correlation ($r = .513$). While the value of this statistic is the same as Pearson's r if you conducted it on the data without the outliers, r_{skip} is a different statistic from Pearson's r and you cannot look up significance in the same way. You first calculate a t_{skip} value with n being the total number in the sample (including the outliers) with the same equation as used with r:

$$t_{skip} = \frac{r_{skip}\sqrt{n-2}}{\sqrt{1 - r_{skip}^2}}$$

or here:

$$t_{skip} = \frac{.857\sqrt{21-2}}{\sqrt{1 - .86^2}} = 7.25$$

This is within rounding error of the value produced above by the **scor** function. This statistic does *not* have the same critical values at *the t* statistic. Wilcox (2003b) ran some simulations and came up with an equation for the critical value for $\alpha = .05$. The equation is:

$$tskip_{critical} = \frac{6.947}{n} + 2.3197$$

which for this example yields $6.947/21 + 2.3197 = 2.65$. The observed value exceeds this, so the skipped correlation is statistically significant at $\alpha = .05$. Wilcox's **scor** function, written for R, produces all these values so you do not have to.

Exercise 16 – Food and drink intake in animals

- Data: Made-up data by Anscombe (1973).
- Library: **MASS, lattice**.

---

[3]Although Spearman's and Pearson's produce the same value with all the data, they will not when the outliers are removed because the data are no longer ranked.

- Research question: What is the relationship between food and drink intake for different types of animal?
- Purpose: To illustrate the **rlm** function, to work with the **split** function, and to stress the importance of graphing your data.

The Anscombe (1973) data set is important in the history of exploratory data analysis (EDA) because it beautifully illustrates how somebody who does not make friends with their data can reach very silly conclusions. We use these data to illustrate robust regression and in particular show when we would expect these methods to make a difference and when we would expect them not to make a difference.

Robust methods are an incredibly active area of statistical research. Methodologists agree that robust methods, of some type, should usually be used. Because of this importance it is tempting to have several chapters on this. But, because of its importance it also means the people writing many of these functions have made them user-friendly. After one of us had a lecture course where about a third of the time was spent on robust methods, the student feedback was: 'shorten that section, all you do is replace **lm** with **rlm**.' This was frustrating because at one level robust methods are so important, and the mathematics behind them are detailed, that they *deserve* much explanation. However, at the pragmatic level of the typical psychology-user this feedback was right, so we will adopt the pragmatic brief approach. This does not mean robust regression is unimportant. Please try **rlm** on your own data!

To make this example more concrete, assume that data are food and drink intake for birds, mammals, insects, and reptiles. The food and drink are in some measure that takes into account body weight. They are stored on the book's web page.

```
webreg <- "http://www.sagepub.co.uk//wrightandlondon//"
anscombe <- read.table(paste(webreg,"animal.dat",
    sep=""),header=T)
attach(anscombe)
```

If you type **anscombe** the data are printed in several long columns. To make them readable on a single screen (assuming the window is open wide enough) type:

```
cbind(anscombe[1:11,], anscombe[12:22,],
    anscombe[23:33,], anscombe[34:44,])
```

	Food	Drink	Type	Food	Drink	Type	Food	Drink	Type	Food	Drink	Type
1	10	8.04	bird	10	9.14	mammal	10	7.46	insect	8	6.58	reptile
2	8	6.95	bird	8	8.14	mammal	8	6.77	insect	8	5.76	reptile
3	13	7.58	bird	13	8.74	mammal	13	12.74	insect	8	7.71	reptile
4	9	8.81	bird	9	8.77	mammal	9	7.11	insect	8	8.84	reptile
5	11	8.33	bird	11	9.26	mammal	11	7.81	insect	8	8.47	reptile
6	14	9.96	bird	14	8.10	mammal	14	8.84	insect	8	7.04	reptile
7	6	7.24	bird	6	6.13	mammal	6	6.08	insect	8	5.25	reptile
8	4	4.26	bird	4	3.10	mammal	4	5.39	insect	19	12.50	reptile
9	12	10.84	bird	12	9.13	mammal	12	8.15	insect	8	5.56	reptile
10	7	4.82	bird	7	7.26	mammal	7	6.42	insect	8	7.91	reptile
11	5	5.68	bird	5	4.74	mammal	5	5.73	insect	8	6.89	reptile

Pretend that you thought Tukey (1977) was misguided when he suggested that graphing data was an integral part of analysis, that you thought Tufte's books (e.g., 2001) on making clear and informative graphs are boring, and that you thought Murrell (2006) had wasted his time writing lots of the graphing procedures in R. If you wanted to see if the animal types were different, you might compare means and standard deviations. To do this use the **tapply** function to calculate the means and standard deviations for the different groups. So, **tapply(Food,Type,mean)** calculates the **mean** of **Food** for each value of **Type**. The command below does the mean and standard deviation (**sd**) for food and drink for each animal type. The **row.names** function adds names to each row for the object **meansd**.

```
meansd <- rbind(tapply(Food,Type,mean),tapply(Food,Type,sd),
    tapply(Drink,Type,mean),tapply(Drink,Type,sd))
row.names(meansd) <- c("Food mean","Food sd","Drink mean",
    "Drink sd")
meansd
```

	bird	insect	mammal	reptile
Food mean	9.000000	9.000000	9.000000	9.000000
Food sd	3.316625	3.316625	3.316625	3.316625
Drink mean	7.500909	7.500000	7.500909	7.500909
Drink sd	2.031568	2.030424	2.031657	2.030579

Having seen these values it would clearly be misguided to run ANOVAs (because the means are all the same so the F values should be 0 or very near zero), but given your views on Tukey, Tufte, and Murrell you may still want to:

summary(aov(Food~Type))

	Df	Sum Sq	Mean Sq	F value	Pr(>F)
Type	3	6.098e-31	2.033e-31	1.848e-32	1
Residuals	40	440	11		

summary(aov(Drink~Type))

	Df	Sum Sq	Mean Sq	F value	Pr(>F)
Type	3	6.818e-06	2.273e-06	5.509e-07	1
Residuals	40	165.008	4.125		

1.848e-32 is .00000000000000000000000000000001848, which is (very) close to zero. Since the research question was to look at the relationship between food and drink intake you might run correlations/regressions for the four groups. To do this you need to use the **split** function to separate the values for the different animal types. A **for** loop is used to run the same analyses on each group.

```
Foodg <- split(Food,Type)
Drinkg <- split(Drink,Type)
for (i in 1:4) {print(paste("Group =",names(Drinkg)[i]))
    print(summary(lm(Drinkg[[i]]~Foodg[[i]])))}
```

```
[1] "Group = bird"

Call:
lm(formula = Drinkg[[i]] ~ Foodg[[i]])

Residuals:
    Min       1Q    Median      3Q      Max
-1.92127  -0.45577  -0.04136  0.70941  1.83882

Coefficients:
             Estimate  Std. Error  t value  Pr(>|t|)
(Intercept)   3.0001      1.1247    2.667   0.02573 *
Foodg[[i]]    0.5001      0.1179    4.241   0.00217 **
---
Signif. codes: 0 '***' 0.001 '**' 0.01 '*' 0.05 '.' 0.1 ' ' 1

Residual standard error: 1.237 on 9 degrees of freedom
Multiple R-Squared: 0.6665,      Adjusted R-squared: 0.6295
F-statistic: 17.99 on 1 and 9 DF, p-value: 0.002170


[1] "Group = insect"

Call:
lm(formula = Drinkg[[i]] ~ Foodg[[i]])

Residuals:
    Min      1Q   Median      3Q     Max
-1.1586  -0.6146  -0.2303  0.1540  3.2411

Coefficients:
             Estimate  Std. Error  t value  Pr(>|t|)
(Intercept)   3.0025      1.1245    2.670   0.02562 *
Foodg[[i]]    0.4997      0.1179    4.239   0.00218 **
---
Signif. codes: 0 '***' 0.001 '**' 0.01 '*' 0.05 '.' 0.1 ' ' 1

Residual standard error: 1.236 on 9 degrees of freedom
Multiple R-Squared: 0.6663,      Adjusted R-squared: 0.6292
F-statistic: 17.97 on 1 and 9 DF, p-value: 0.002176


[1] "Group = mammal"

Call:
lm(formula = Drinkg[[i]] ~ Foodg[[i]])

Residuals:
    Min      1Q   Median      3Q     Max
-1.9009  -0.7609  0.1291  0.9491  1.2691

Coefficients:
             Estimate  Std. Error  t value  Pr(>|t|)
(Intercept)   3.001       1.125     2.667   0.02576 *
Foodg[[i]]    0.500       0.118     4.239   0.00218 **
---
Signif. codes: 0 '***' 0.001 '**' 0.01 '*' 0.05 '.' 0.1 ' ' 1
```

```
Residual standard error: 1.237 on 9 degrees of freedom
Multiple R-Squared: 0.6662,      Adjusted R-squared: 0.6292
F-statistic: 17.97 on 1 and 9 DF, p-value: 0.002179


[1] "Group = reptile"

Call:
lm(formula = Drinkg[[i]] ~ Foodg[[i]])


Residuals:
       Min         1Q      Median         3Q         Max
-1.751e+00  -8.310e-01  1.258e-16  8.090e-01  1.839e+00


Coefficients:
             Estimate  Std. Error  t value  Pr(>|t|)
(Intercept)    3.0017      1.1239    2.671   0.02559 *
Foodg[[i]]     0.4999      0.1178    4.243   0.00216 **
---
Signif. codes: 0 '***' 0.001 '**' 0.01 '*' 0.05 '.' 0.1 ' ' 1


Residual standard error: 1.236 on 9 degrees of freedom
Multiple R-Squared: 0.6667,      Adjusted R-squared: 0.6297
F-statistic: 18 on 1 and 9 DF, p-value: 0.002165
```

There is nothing in this numeric output to show that the relationships are different; most of the key statistics that people look at are the same for each of the four groups. Of course, some of you may be thinking that the Tukey-Tufte-Murrell approach may have some merit, so you decide to look at the scatterplots for the different animal types. There is a package called **lattice** (Sarkar, 2008), also used in Chapters 4 and 8, that can do this type of graph quickly. This package is useful and is described in detail in Murrell (2006). However, in this book we have relied mostly on the traditional graph methods since it would take many pages to scratch the surface of **lattice**'s capabilities. But, just as an example the following procedures and Figure 9.5.

```
library(lattice)
xyplot(Drink~Food|Type)
```

A similar plot using traditional graphs (i.e., the **plot** function) can be made. The **for (i in 1:4)** has R make the **plot** for each of the four groups. **main=(paste(names(Foodg)[i]))** is used to put the names of the groups in above the individual scatterplots in Figure 9.6.

```
par(mfrow=c(2,2))
for (i in 1:4) {
   x <- lm(Drinkg[[i]]~Foodg[[i]])
   plot(Foodg[[i]],Drinkg[[i]],main=
       (paste(names(Foodg)[i])),xlab="Food intake",
       ylab="Drink intake",ylim=c(0,15),xlim=c(0,20))
   abline(x)}
par(mfrow=c(1,1))
```

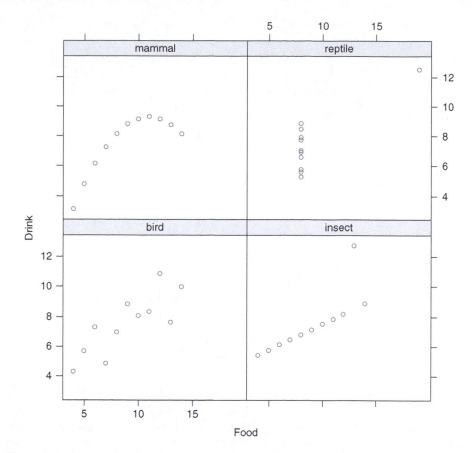

Figure 9.5 Scatterplots for the four animal types comparing drink intake with food intake. The graph is made with **xyplot** from the **lattice** library and the data were made by Anscombe (1973)

Two things are obvious from these data. First, the relationship between food and drink intake is very different for the different types of animals. Second, these data are meticulously made-up so that they have the same means, standard deviations, correlations, regression lines, but the relationships are all different. Clearly if someone had just reported the numeric statistics they would have reached the wrong conclusions. For those of you who teach undergraduate statistics, we encourage you to use these data to illustrate the need for graphing. Here we use them for robust regression.

There are several choices of robust regressions. The main way that the regressions vary is according to the loss function. Earlier in the chapter we talked about least absolute values, trimming, and M-Estimators. M-estimators are the most common and these are what are built within the **rlm** function. The **psi** option allows different loss functions and at present it allows ones developed by Huber, Hampel, and Tukey. Each of these has certain values you can tell the computer to use to specify it. While there are arguments among statisticians about the best of these, within psychology we feel trying multiple methods until you get a model you like would be wrong here (we are not in general against this approach, see the conclusions of model selection in Chapter 5) and we feel

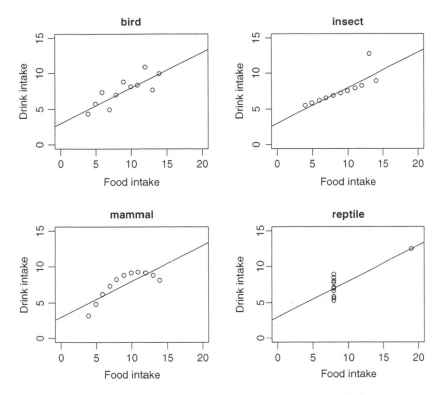

Figure 9.6 The scatterplots from Figure 9.5 remade with the **plot** command. The standard linear regression is added to each plot

there are no psychology examples which would lend themselves better to one of these than another. Therefore, we recommend only running the default (which is the Huber function).

The question is, how will a robust regression help in modeling the Anscombe data? For the birds, the data look as if they conform, roughly, to the assumptions of least squares regression, so we would not expect large differences from a robust regression. Like we said before, for the user the **rlm** function works in a very similar way to the **lm** function. It is part of the **MASS** library (Venables & Ripley, 2002), a large collection of functions for R (and S-Plus) that are used by most R users. The results below are very similar to those found with the **lm** procedure. Unlike the **lm** procedure, the output does not give you the p values associated with the coefficients. If you find p values useful, you need to take the t value from the output and the degrees of freedom (calculate this from $df = n - 2 = 9$) and use the **pt** function. This is for the probability of a t value. For this example the probability associated with the food coefficient for birds type: **pt(3.3351,9,lower.tail=F)*2**. It produces $p = 0.009$ so you can write $t(9) = 3.34, p < .001$.

```
library(MASS)
birds <- rlm(Drinkg[[1]]~Foodg[[1]])
summary(birds)
```

```
Call: rlm(formula = Drinkg[[1]] ~Foodg[[1]])
Residuals:
      Min         1Q     Median        3Q       Max
  -1.98298   -0.48442   -0.08247   0.69278   1.78312


Coefficients:
                Value   Std. Error   t value
(Intercept)   2.9837   1.4476        2.0611
Foodg[[1]]    0.5061   0.1518        3.3351


Residual standard error: 1.109 on 9 degrees of freedom


Correlation of Coefficients:
              (Intercept)
Foodg[[1]]   -0.9435
```

Class *Insecta*! There are over 2,000 species of praying mantises! Their scatterplots in Figures 9.5 and 9.6 show a straight line of data with one point off the line. This point will be influential for the standard regression, but it will make less of an impact for robust regressions. The regression line does change, but the most noticeable difference in the output is that the *t* values have rocketed upwards. This makes sense. Because all the others are on a straight line, if the regression is lessening the weight of the outlier, then the *p* values drop accordingly.

```
insect <- rlm(Drinkg[[2]]~Foodg[[2]])
summary(insect)
```

```
Call: rlm(formula = Drinkg[[2]] ~ Foodg[[2]])
Residuals:
       Min          1Q      Median         3Q        Max
  -0.004996   -0.002859   -0.000722   0.002867   4.242101


Coefficients:
                Value   Std. Error   t value
(Intercept)   4.0035   0.0040       990.3355
Foodg[[2]]    0.3457   0.0004       815.8284


Residual standard error: 0.005248 on 9 degrees of freedom


Correlation of Coefficients:
              (Intercept)
Foodg[[2]]   -0.9435
```

The third group is mammals. Clearly a linear regression is wrong because of the curved pattern. But the robust regressions that we are describing are still linear. There are not any large outliers, so we would not expect, and do not find, many differences between the robust regression and the standard one.

```
mammals <- rlm(Drinkg[[3]]~Foodg[[3]])
summary(mammals)

Call: rlm(formula = Drinkg[[3]] ~Foodg[[3]])
Residuals:
     Min          1Q    Median          3Q         Max
-1.95997    -0.81997   0.07003    0.89003    1.21003

Coefficients:
              Value    Std. Error   t value
(Intercept)  3.0600   1.3113        2.3336
Foodg[[3]]   0.5000   0.1375        3.6375

Residual standard error: 1.216 on 9 degrees of freedom

Correlation of Coefficients:
            (Intercept)
Foodg[[3]]  -0.9435
```

What is needed here is a term in the regression for the curve, so we tried a quadratic regression. From this it is clear we have uncovered how Anscombe (1973) created the data for this group. A simple quadratic fits these data nearly perfectly.

```
mammals2 <- lm(Drinkg[[3]] ~ poly(Foodg[[3]],2))
summary(mammals2)

Call:
lm(formula = Drinkg[[3]] ~ poly(Foodg[[3]], 2))

Residuals:
      Min          1Q      Median          3Q         Max
-0.0013287   -0.0011888   -0.0006294   0.0008741   0.0023776

Coefficients:
                         Estimate  Std. Error  t value  Pr(>|t|)
(Intercept)             7.5009091  0.0005043    14875   <2e-16 ***
poly(Foodg[[3]], 2)1    5.2440442  0.0016725     3135   <2e-16 ***
poly(Foodg[[3]], 2)2   -3.7116396  0.0016725    -2219   <2e-16 ***
---
Signif. codes:  0 '***' 0.001 '**' 0.01 '*' 0.05 '.' 0.1 ' ' 1

Residual standard error: 0.001672 on 8 degrees of freedom
Multiple R-Squared:       1,        Adjusted R-squared: 1
F-statistic: 7.378e+06 on 2 and 8 DF, p-value: < 2.2e-16
```

The final group is reptiles. Like the insects, there is one data point that does not fit the pattern. However, the pattern for the others is to have the exact same values for food intake. If we removed the outlying datum, the standard deviation for the remainder of this group would be 0, and therefore without the outlier there is no correlation or regression (not a correlation of 0, no correlation). The odd point is a univariate outlier, but it actually is a very influential point for the standard regression. Thus, we would expect (and we find) it would still be weighted highly by this robust regression.

```
reptiles <- rlm(Drinkg[[4]]~Foodg[[4]])
summary(reptiles)

Call: rlm(formula = Drinkg[[4]] ~ Foodg[[4]])
Residuals:
        Min          1Q      Median          3Q         Max
-1.749e+00  -8.286e-01  -1.776e-15   8.114e-01   1.841e+00

Coefficients:
             Value   Std. Error  t value
(Intercept)  2.9976  1.2570      2.3847
Foodg[[4]]   0.5001  0.1318      3.7955

Residual standard error: 1.351 on 9 degrees of freedom

Correlation of Coefficients:
             (Intercept)
Foodg[[4]]   -0.9435
```

Of course this is an odd set.[4] It is worth considering how the **scor** function would work with the data for each of these animal types. For the birds, **scor** did not remove any outliers so produced Pearson's r (.82). For insects, it does not count the one wayward point as an outlier, so also produces Pearson's r. For the mammals, it identifies one point at the end as an outlier. Because this is near the regression line, when it is excluded the correlation value drops to .76. For the reptiles, **scor** produces an error because the outlier is excluded and this leaves no standard deviation in food intake. The graphs though tell the story. You can add a small jitter to these points **scor(Drinkg[[4]],jitter(Foodg[[4]]))**) to allow a correlation to be calculated, but its value will be near zero since it will just compare the drink variable with random noise.

SUMMARY

Robust procedures are recommended by the APA (Wilkinson et al., 1999) for analyzing data and they increase the chances that you find significant effects. Therefore, if you want a significant effect (and a lot of psychologists do), use these. If a reviewer complains that they think you are using some fancy statistic to squeeze out a significant result, then point them to the APA report (or Wilcox's books, or this book, or lots of places) as justification.

While detailed descriptions of different estimators could have been given, the approach taken here was both simpler and with an education purpose. We began with Spearman's ρ because it is very popular. We wanted to stress the difficulty once you have a statistic based

---

[4]Another data set similar to the reptiles can be made with **drink <- c(rep(1,10),k)** and **food <- 1:11**. Let **k** be different values and look at the correlation. It will be $-.5$, 0, or $+.5$. The size of **k** does not affect for the correlation for these data, only which side of 1 that it is on.

on ranks trying to make sensible quantitative statements in terms of the raw data. There are more advanced rank based procedures, but they are less popular than other modern alternatives. Because of this we recommend using a numerical transformation where possible (or using GLMs if appropriate) because the values can be back-transformed onto the original scales. Both the main rank-based procedures and the standard transformations work only on univariate outliers. The second example showed a conceptually simple method for calculating a correlation that excludes bivariate outliers. The procedure works in three steps: a) determine which values are outliers and remove these; b) calculate the correlation on the remaining items; and c) calculate the p value associated with this correlation. Showing this procedure also allowed us to introduce readers to Wilcox's functions. The **scor** function is relatively new, but it has promise.

In practice, the most useful approach uses the **rlm** function, so we ended with this. Like **scor** it lessens the impact of bivariate outliers because these have the largest residuals, but unlike **scor** it only lessens their impact if they are outliers away from the regression line. This was seen with the reptiles in the final example, where an outlier near the regression line was still highly influential. When describing robust methods it is tempting to say to use them for everything and that you will no longer need to worry about outliers or any other oddities in the data. It may seem as if they are a panacea for everything. This is not true and is why we chose the Anscombe (1973) data sets. Looking at the scatterplots, it is clear that the standard linear regression is only appropriate for one of the data sets (birds), but the robust regression only addresses the odd pattern in one of the remaining sets (insects), where there is a large residual. Therefore, as Anscombe intended, his data sets show the importance of looking at your data graphically before deciding which numeric procedure to apply.

SOME WORDS/CONCEPTS WORTH REMEMBERING

R concepts

* bootstrap methods with a bivariate statistics (Spearman's ρ).

R functions

* **scor**: skipped correlation coefficient;
* **rlm**: robust linear model;
* **pt**: probability associated with a t value.

Statistical concepts

* loss function: what is minimized when estimating a regression;
* trimmed: removing the tails of the distribution;
* outliers and inference: outliers decrease the chance of significance;
* to Tommy Franks: to ridicule the Geneva Convention*;
* M-estimators: a popular method of robust estimation;
* Anscombe's data: a neat data set showing the need for graphs.

* According to a quotation from the *Lancet* paper.

FURTHER READING

Wilcox, R. R. (1998). How many discoveries have been lost by ignoring modern statistical methods? *American Psychologist*, *53*, 300–314. This article is written for a general psychology audience and goes through the reasons why robust methods should be used.

Wilcox, R. R. (2003a). *Applying contemporary statistical techniques*. Orlando, FL: Academic Press. Wilcox has written several books at different levels of complexity. This one is an introductory statistics book, written for intelligent people, but who have not taken much statistics.

10

Conclusion – make your data cool

Learning outcomes

1. Regression towards the median.
2. To consider the discovery and dissemination aspects of statistics.

The Oxford English Dictionary lists several definitions for regression. The most used is to 'go back.' One psychology use of this is with hypnotic age regression, which means mentally revisiting your childhood. In some faiths it can mean going back to previous lives on distant worlds (probably the most discussed of these is Scientology, http://sf.irk.ru/www/ot3/otiii-gif.html, but other faiths have beliefs about things like life-after-death). Within data analysis regression takes on two related but distinct meanings.

Historically, the first statistical use of regression is 'regression towards the mean/median' which follows the English definition of going back. This was introduced by Francis Galton (1886), one of the most influential (and controversial) people in early psychology (Brookes, 2004). Galton had noticed in studies of seeds that small seeds tended to produce seeds that were slightly larger than themselves, and that large seeds tended to produce seeds that were slightly smaller than themselves. There was a tendency for the offspring to move towards middle sized seeds. He found the same thing with humans, and used this to argue for a theory of heredity (and his views on eugenics). Given the topic of this book, his paper is of enough historical importance that it is worth further discussion.

In his Table 1 Galton lists the average height (in inches) of 205 parents and their 928 adult offspring (he adjusts female heights so that there is not a gender effect and describes numerous methodological issues; see also Wachsmuth et al., 2003). To demonstrate the concept of regression towards (not 'to' since you would still expect the offspring of tall parents to be taller than) the middle, we will create some data. To simplify things, we look at the simpler (and less enjoyable) case of asexual reproduction where the offspring receives 100% of a single parent's genes. There would be some genetic basis for height and some non-genetic basis. An individual's height will be determined by the genetic and the non-genetic causes. Since we created the data ourselves, we know what proportion of variation we would expect from both of these causes in this sample. We would expect people's height based just on genetic factors to be 70 ± 10 inches and then the environmental factors can alter this by ± 10 inches. So genetic factors and environmental factors should have about equal impacts.

The code below makes these data. We use **set.seed(47)** so that the data can be recreated. **runif(100,60,80)** creates **100** cases from a variable with a uniform distribution between **60** and **80**. We are assuming that there is no correlation between a parent's and her child's environment, so pretend this is part of some bizarre experiment.

```
set.seed(47)
genoheight <- runif(100,60,80)
parentheight <- genoheight + runif(100,-10,10)
kidheight <- genoheight + runif(100,-10,10)
```

Below is the code for Figure 10.1, using a variety of functions that we have been using before (**abline(0,1)** means a line with a 0 intercept and a slope of one; **abline(v=X)** and **abline(h=X)** draw vertical and horizontal lines at the value **X**; and **\n** within the **text** command means a line break).

```
plot(parentheight,kidheight,pch=20,xlab="Parent's height in
    inches",ylab="Child's height in inches")
abline(0,1)
abline(h=mean(kidheight),v=mean(parentheight),lty="dashed")
text(55,84,"Most kids of short\nparents above\nthe diagonal",
    pos=4)
```

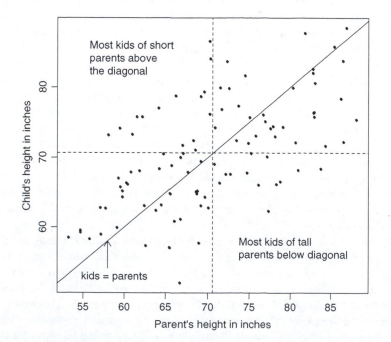

Figure 10.1 Regression towards the median

```
text(73,57,"Most kids of tall\nparents below diagonal", pos=4)
arrows(58,54,58,58,length=.1)
text(54,53, "kids = parents",pos=4)
```

The correlation and its confidence interval can be found with the **cor.test** function:

cor.test(parentheight,kidheight)

```
Pearson's product-moment correlation

data: parentheight and kidheight
t = 7.0618, df = 98, p-value = 2.394e-10
alternative hypothesis: true correlation is not equal to 0
95 percent confidence interval:
  0.4337967 0.6975807
sample estimates:
  cor
0.5807332
```

Figure 10.1 and the statistical output show that the heights of parents with their children are related ($r = .58$, with a 95% CI including .5, which is expected because the variable is half from the genotype variable and half an environmental/random variable), but that kids of short parents tend to be taller than their parents (dots above the diagonal) and that kids of tall parents tend to be shorter than their parents (dots below the diagonal). This regression towards the mean, median or mediocrity (as it gets labeled in different places) is now taught as an methodological artifact to avoid, but in Galton's day it was used for supporting his theory. Today, the most important thing to take from it is that most measures are based both on some real value and some measurement error (in behavioral genetics the non-genetic component includes error), and therefore if you sample just on the basis of the observed value then you would expect later measures to shift towards the middle of the distribution. An example is if an educational researcher was interested in an intervention to improve poor students' scores. If the researcher gave a test to students and then gave the intervention to the half of students who performed the worse on this test, it is likely that at re-test this group would perform better then they had, and that the other group would perform worse than they had. This, however, could just be the methodological artifact of regression towards the median.

In the history of regression Galton's paper was also important for drawing the regression equation line. The most common use of the word 'regression' within contemporary data analysis/statistics is as a method to create expected values from some model and to compare these with observed values. This is the use of 'regression' assumed in this book. The basic elements of a regression are: a set of observed values on the left side of the equal sign, an equal sign, a model based on observed values of other variables on the right side of the equal sign, and an error term. Because most statistics can be framed in this way means that most statistics can be described as regressions. Of course, just being able to rename an ANOVA as a special case of regression does not increase our understanding of either, but as shown in Chapter 3, being able to show their equivalence and then build on each does. Further, the advances in each of the other chapters also help to increase our understanding of statistics. The purpose of this book is not, however, to increase your understanding of statistics for the sake of it, but so that you can use these techniques for understanding your data.

Statistics has two components: discovering/evaluating patterns in nature and describing these to an audience. The history of statistics is of a lot of intelligent and insightful people describing ways to turn your complex and seemingly chaotic data into information that illuminates your theories of nature. Because this book is on the art of regression and therefore it is really on the art of all statistics, it is fitting that we end with two of our favorite philosophies of this discovering/disseminating duality of statistics.

The first philosophy is based on many people's work, but most notably Tukey (1977), and that is we should focus on the data, or in Rosenthal's terms we should make friends with our data (Wright, 2003). The second philosophy is based on Abelson's MAGIC (1995) criteria for persuading people about the story your data have to tell. The first philosophy is about turning data into information, and the second is about using this information to persuade the reader about the hypotheses under consideration.

The 'making friends with data' component is best exemplified by Tukey and colleagues when discussing exploratory data analysis (EDA). Hoaglin et al. (2000) discussed the four themes, or four Rs, of EDA: resistance, residuals, re-expression, and revelation.

- Resistance: statistics should not be greatly affected by a few points,
- Residuals: do some points appear to have arisen from different processes than the others,
- Re-expression: sometimes the original variables need to be transformed,
- Reveal: what do your data tell you?

These four Rs should be considered whenever you are trying to extract coherent patterns from data. The hope is that the techniques presented throughout this book will help you address each of these four Rs.

Figure 10.2 shows how you would use this approach to evaluate a regression. If you only ask whether the p value is significant, this would be bad. It is worth remembering that finding a significant p just tells you that your sample size was large enough to detect the effect. If you only use the p value to evaluate a model you get a ☹. You should also look at the size of the effect, and with regressions this is often Pearson's r or some related statistic. This is better, but we know that r is just a single number and it can be influenced by outliers and other influential points. You should ask whether the model is robust. This may involve running robust methods, but it can also be addressed by seeing if your conclusions would be very different if you just moved a couple of points around. If your conclusions are dependent on the location of just a couple of points, then you should be cautious in making these conclusions. As the Anscombe (1973) data sets in Chapter 9 illustrate, it is vital to look at your data. These are the first four of the five

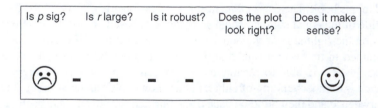

Figure 10.2   Ways to assess the value of a model (i.e., to make friends with your data)

steps in Figure 10.2. You may find a significant *p* value, a large *r* value, have tested that this is robust, and have looked at the plots. But, if your conclusion implies that apples do not fall to Earth with gravity, your conclusion is wrong. The final step to achieve the coveted ☺ requires that your finding fits into a coherent view of science. This leads to the second philosophy of discovery/dissemination of statistics.

The second philosophy describes what to do with the information that has carefully been considered with the four Rs and evaluated with Figure 10.2. Abelson (1995) argued that the general rules of communication and persuasion should be applied to describing this information within results sections (see also Wright & Williams, 2003). He described the MAGIC criteria for a good results section. MAGIC stands for: Magnitude, Articulation, Generality, Interest, and Credibility. These labels are fairly self-explanatory:

- Magnitude: report effect sizes,
- Articulation: focus the reader on the effects of importance,
- Generality: state how broadly your results should generalize,
- Interest: make the reader excited (which means you should be excited),
- Credibility: provide good evidence for your claims.

Abelson's MAGIC can be applied to these observed patterns to produce an accurate, convincing and clear story. MAGIC is a great acronym because it stresses the amazement that your readers should experience when reading your results section. One of our colleagues was giving a talk once and before showing his results he said: 'these data are really cool!' His excitement spread through the audience. Our final advice is to: Make your data cool!

SOME WORDS/CONCEPTS WORTH REMEMBERING

R functions

- `runif`: for creating uniformly distributed random numbers.

Statistical concepts

- regressions towards the median: scores are based on a true score and error.
- 4 Rs of statistics : Hoaglin et al.'s approach to statistics.
- MAGIC: Abelson's description for making statistics useful.
- smile scale: our way of summarizing an approach to statistics.

FURTHER READING

Abelson, R. P. (1995). *Statistics as principled argument*. Mahwah, NJ: Lawrence Erlbaum Associates. This is the ideal summer reading for anybody who plans to teach their first undergraduate psych-stats course. It allows you to step back from the algorithms and think about why you are doing statistics.

Glossary of R functions used in this book

The following is an alphabetical list of most of the R functions and packages used in this book, as well as some of the important options.

<-	to assign objects to each other **x <- 6+4** assigns 10 to x
->	to assign objects to each other **6+4 -> x** assigns 10 to x
?	for help with R functions when you know the function name
#	use when you want R to ignore the line (for comments)
$	to access part of an object
+ - * /	addition, subtraction, multiplication, division of two numbers
^	**x^y** is x to the power y (****** also works but not used in this book)
~	connects the response variable and the model in all regressions
:	**x:y** is the sequence from x to y if they are numbers
:	**x:y** is the interaction between variables x and y in a regression
*	**x*y** is the interaction and all nested effects of variables x and y in a regression
/n	puts a line break within **text** commands
abline	for adding straight lines to plots
aggregate	used to analyze or report statistics at the group level
anova	produces the ANOVA table with sum of squares for many regression objects (and **aov** objects) and compares them
aov	an ANOVA test to compare means
arrows	adds arrows to a plot
as.data.frame	reads a set of variables as a data frame
as.factor	reads a variable as a factor (i.e., categorical)
as.numeric	reads a variable as numeric (i.e., numbers)
as.ordered	reads a variable as ordinal
axes	an option within **plot**. Can take several values including **FALSE**.
attach	used to make a data set active in your environment
boot	a package for bootstrapping, and also the name of the main function for bootstrapping
boot.ci	calculates several bootstrap confidence intervals

box	draws a box around a plot
breaks	used in histograms and elsewhere to show where to split a numeric variable into bins
bs	B-Splines function from the **splines** package
c	concatenate. Puts items together
car	a package with several useful functions for regression
cbind	binds columns together
cex	change the size of different parts of a plot
chisq.test	the two variable chi-square test of no association
coef	used to access the coefficients of many regression objects
col	changes the color of many parts of a graph
contrasts	the contrasts used for categorical predictor variables. This function can be used to show the contrasts and to set them
cor	for a correlation/covariance matrix of several variables
cor.test	for a correlation between pairs of variables
cut	to split a variable into categories. Uses the option **break** or **br**
cv.lars	used for cross validation of **lars** models in order to decide how much shrinkage is appropriate
demo	runs demonstrations of some R functions. Not available for most, but **demo(plotmath)** is described in the text. **example** is also available
diag	the diagonal of a matrix
digits	an option in **print**, **format** and elsewhere to control the number of digits printed
dim	to show or to set the dimensions of an object. For example, **x <- 1:12** **dim(x) <- c(4,3)** makes a 4 x 3 structure
dpois	to make a Poisson distribution
e1071	a package with several functions including one for skewness
each	an option in **seq** to control whether an entire set is repeated or whether each element is repeated
exp	**exp(x)** is e^x
degree	an option within **poly** and **bs** for the degree polynomial to use
df	an option within **bs** for the complexity of the spline
detach	to remove a data set from the active environment
expression	used to type odd symbols and mathematical characters
FALSE	a value of a Boolean variable. **F** also works
fBasics	a package with several functions including one for skewness
fitted.values	used to find predicted values. **predict** can also be used
for	one of the basic ways in R to repeat a sequence of commands
foreign	a package which allows data to be read and to be written to other statistics packages
format	used to control the format of output, usually because the default format has too many digits
function	tells R that the argument is a function

`gam`	a package with functions for generalized additive models. Also the name of the basic generalized additive model (GAM) function
`glm`	for generalized linear models. The **family** option determines which error distribution is assumed, and also has the usual link function for each distribution as default
`header`	an option in **read.table** that means the first line of the file is variable names
`help`	for help with R functions when you know the function name
`help.search`	to search for a phrase in the R files
`hist`	the histogram
`install.packages`	to load packages from the web (the Comprehensive R Archive Network, or CRAN) onto your computer
`intervals`	The confidence intervals of **lme** and **nlme** objects
`invisible`	tells R not to print the value of a function. Useful if writing your own functions
`jitter`	adds a small random amount to each value. Useful for scatterplots when discrete data are plotted
`knots`	an option with **bs** for the location and/or number of knots
`ks.test`	Kolmogorov-Smirnov test for a distribution's shape
`labels`	used to assign labels and to find out the labels of a factor
`lars`	the least angle regression package and function. This solves the lasso efficiently, as well as a few other regression techniques
`lattice`	a package for trellis graphs. Only briefly covered in this book. See Murrell (2006) for more details
`leaps`	a package that includes best subset regression. Also a function to calculate the best subset. How 'best' is defined is dependent on the **option** method used
`legend`	adds a legend to plot and is an option in some other functions
`lend`	controls how lines connect. Only necessary to worry about if using thick lines
`length`	how long a variable is
`library`	to load packages into your active environment so they can be used
`lines`	used to draw lines, including curved ones, onto a plot. More flexible than the **abline** function
`list`	a type of data object
`lm`	linear model (for simple and multiple regression)
`lm.object`	the regression object created from **lm**
`lm.ridge`	for ridge regression. Included in the **MASS** package
`lme`	for linear mixed models
`lmer`	for generalized linear mixed models
`log`	the logarithm. The default is the natural logarithm
`lower.tri`	the lower triangle of a matrix
`lty`	changes the line style
`lwd`	changes the line width in a plot

main	the title given within the different plotting functions
MASS	Venables and Ripley's package which is used for many procedures. Their book is valuable too
mean	the mean and trimmed mean
median	the median
mgcv	a package with functions for complex generalized additive models
names	to reveal the names in a data set
nlme	non-linear mixed effect models. These are not covered in this book
options	to show or to change various aspect of your R environment
order	used to order the cases within a data set according to one of the variables
par	to change parameters for graphs
par(mar=	to change the margin on a graph
par(mfrow=	to print multiple graphs in a grid format
paste	combines pieces of text
pch	controls the style of the points in a scatterplot
pchisq	the p value associated with the χ^2 distribution. Depending on your needs you may wish to set **lower.tail=FALSE**
pcr	Principal components regression. Conducts PCA on data then uses these components as predictors in a regression
plot	the basic plot function. Has many options and works differently depending on what kind of object is entered into it
pls	A package with **pcr** and **plsr** (and functions for multivariate regression)
plsr	partial least squares regression. Simultaneously tries to find variables that fit together and predict the response variable
points	to add points to a plot
poly	makes orthogonal polynomial contrasts. Useful for polynomial regressions
pos	controls position of text
predict	calculates the predicted values from a regression model either for the observed data or for a new data file
princomp	runs a principal components analysis
psi	To choose the loss function options for **rlm**
pt	probability of a t value
quantile	finds the quantiles of any variables
rank	ranks the values of a variable
rbind	binds rows together
rbinom	to create binomially distributed random variables
read.spss	reads data from an SPSS data file from your computer (at present does not read SPSS files from the web)
read.table	reads data in a text format from a file on your computer or on the web
rect	draws a rectangle onto a graph
regsubsets	finds best fitting regression for different numbers of predictor variables. How 'best fitting' is decided is based on the **scale** option

rep	to repeat a value, or set of values, a specified number of times
rlm	robust linear models from the **MASS** package
rnorm	to create normally distributed random variables
runif	for creating random numbers from a uniform distribution
s	a regression spline used in **gam**, **mgcv**, and elsewhere
scor	Wilcox's skipped correlation function
sd	standard deviation
sep	an option in paste which says what should separate the pieces of text. **sep=" "** means nothing should separate them
set.seed	to set the seed for random number generation. Useful for creating data so that you get the same numbers each time
shapiro.test	Shapiro-Wilks test for normality
show.signif.stars	an option to show or not to show *****s for levels of significance
skewness	calculates skewness. Located in **e1071** and **fBasics**
sort	used to sort a data set according to one variable
source	to load functions from web locations
spline	a smooth line connecting points in a graph
splines	a package for regression splines
split	creates a new object with the values of one variable separated into different parts according to the values of another variable
spm	scatterplot matrix. A function in the **car** package
sqrt	the square root function
summary	reports summary output from most statistical objects
t.test	a t test. Default assumes group variances differ. To assume they are the same include the option **var.equal=TRUE**
table	to construct a contingency table
tapply	to calculate a statistic for different groups for a variable
text	to add text to plots
trunc	truncates the values of a variable
TRUE	a value of a Boolean variable. **T** also works
update	used to slightly change regression models so that you do not have to re-write the entire command
upper.tri	the upper triangle of a matrix
which	returns the index that satisfies the function
which.max	returns the index for the largest value of a variable
which.min	returns the index for the smallest value of a variable
wilcox.test	Wilcoxon tests
write.foreign	to write data to a file in another statistics package (shown for writing to SPSS in the text)
write.table	to write data to a file
xlab	the x axis label

`xlim`	sets the lower and higher values for the x axis
`xy.plot`	multiple scatterplots. A trellis graph from the **lattice** package
`ylab`	the y axis label
`ylim`	sets the lower and higher values for the y axis

The code for running all the analysis is available on the book's web page. This means you do not need to retype code. Also, updates, the data, and links are available on the page. It is: http://www.sagepub.co.uk/wrightandlondon.

References

Abelson, R. P. (1995). *Statistics as principled argument*. Mahwah, NJ: Lawrence Erlbaum Associates.

Agresti, A. (2002). *Categorical data analysis* (2nd ed.). Hoboken, NJ: John Wiley & Sons.

Anscombe, F. J. (1973). Graphs in statistical analysis. *American Statistician, 27,* 17–21.

Ayers, S., Wright, D. B. & Wells, N. (2007). Post-traumatic stress in couples after birth: Association with the couple's relationship and parent-baby bond. *Journal of Reproductive and Infant Psychology, 25,* 40–50.

Banks, W. P. (1970). Signal detection theory and human memory. *Psychological Bulletin, 74,* 81–99.

Barth, J. M., Dunlap, S. I., Dane, H., Lochman, J. E. & Wells, K. C. (2004). Classroom environment influences on aggression, peer relations, and academic focus. *Journal of School Psychology, 42,* 115–133.

Bartholomew, D. J., Steele, F., Moustaki, I. & Galbraith, J. (2002). *The analysis and interpretation of multivariate data for social scientists*. London: Chapman & Hall/CRC.

Bates, D. & Sarkar, D. (2007). *lme4: Linear mixed-effects models using S4 classes.* R package version 0.9975–13.

Berndt, E. R. (1991). *The practice of econometrics*. New York: Addison-Wesley.

Box, G. E. P. & Cox, D. R. (1964). An analysis of transformations. *Journal of the Royal Statistical Society: B, 26,* 211–246.

Brookes, M. (2004). *Extreme measures: The dark visions and bright ideas of Francis Galton*. New York: Bloomsbury Publishing.

Burnham, G., Lafta, R., Doocy, S. & Roberts, L. (2006). Mortality after the 2003 invasion of Iraq: A cross-sectional cluster sample survey. *The Lancet, 368,* 1421–1428.

Canty, A. & Ripley, B. (2008). *boot: Bootstrap R.* (S-Plus) functions. R package version 1.2–33.

Chatfield, C. (2003). *The analysis of time series: An introduction* (6th ed.). Chapman & Hall/CRC.

Cliff, N. & Keats, J. A. (2002). *Ordinal measurement in the behavioral sciences*. Mahwah, NJ: Lawrence Erlbaum Associates.

Clark, H. H. (1973). The language-as-fixed-effect fallacy: A critique of language statistics in psychological research. *Journal of Verbal Learning and Verbal Behavior, 12,* 335–359.

Cohen, J. (1968). Multiple regression as a general data-analytic system. *Psychological Bulletin, 70,* 426–443.

Cohen, J. (1976). Random means random. *Journal of Verbal Learning and Verbal Behavior, 15,* 261–262.

Cohen, J. (1983). The cost of dichotomization. *Applied Psychological Measurement, 7,* 249–253.

Cohen, J. (1988). A power primer. *Psychological Bulletin, 112*, 155–159.

Cohen, J. (1990). Things I have learned (so far). *American Psychologist, 45*, 1304–1312.

Cohen, J. (1994). The Earth is round ($p < .05$). *American Psychologist, 49*, 997–1003.

Collett, D. (2003). *Modelling binary data* (2nd ed.). Boca Raton, FL: Chapman & Hall/CRC.

Conover, W. J. & Iman, R. L. (1981). Rank transformations as a bridge between parametric and nonparametric statistics. *American Statistician, 35*, 124–129.

Cox, D. R. (2006). *Principles of statistical inference.* Cambridge, UK: Cambridge University Press.

Crawley, M. J. (2005). *Statistics: An introduction using R.* Chichester, UK: Wiley.

Crawley, M. J. (2007). *The R book.* Chichester, UK: Wiley.

DeCarlo, L. T. (1998). Signal detection theory and generalized linear models. *Psychological Methods, 3*, 186–205.

DiCiccio, T. J. & Efron B. (1996). Bootstrap confidence intervals (with Discussion). *Statistical Science, 11*, 189–228.

Dienes, Z. (2008). *Scientific and statistical inference: Conceptual issues in psychology.* New York: Palgrave Macmillan.

Dimitriadou, E., Hornik, K., Leisch, F., Meyer, D. & Weingessel, A. (2008). *e1071: Misc Functions of the Department of Statistics (e1071), TU Wien.* R package version 1.5–18.

Efron, B. & Gong, G. (1983). A leisurely look at the bootstrap, the jackknife, and cross-validation. *American Statistician, 37*, 36–48.

Efron, B., Hastie, T., Johnstone, I. & Tibshirani, R. (2004). Least angle regression. *Annals of Statistics, 32*, 407–499.

Embretson, S. E. & Reise, S. P. (2000). *Item response theory for psychologists.* Mahwah, NJ: Lawrence Erlbaum Associates.

Faraway, J. J. (2006). *Extending the linear model with R: Generalized linear, mixed effects and nonparametric regression models.* Boca Raton, FL: Chapman & Hall/CRC.

Festinger, L. & Carlsmith, J. M. (1959). Cognitive consequences of forced compliance. *Journal of Abnormal and Social Psychology, 58*, 203–210.

Field, A. P. (1999). *Discovering statistics using SPSS* (3rd ed.). London: Sage Publications.

Fisher, R. A. (1925). *Statistical methods for research workers.* London: Oliver and Boyd. (Available: http://psychclassics.yorku.ca/Fisher/Methods/)

Fox, J. (2002). *An R and S-Plus companion to applied regression.* Thousand Oaks, CA: Sage Publications.

Fox, J. (2008). *car: Companion to Applied Regression.* R package version 1.2–8.

Fox, J. and with contributions from Ash, M., Boye, T., Calza, S., Chang, A., Grosjean, P., Heiberger, R., Kerns, G. J., Lancelot, R., Lesnoff, M., Messad, S., Maechler, M., Murdoch, D., Neuwirth, E., Putler, D., Ripley, B., Ristic, M. & Wolf, P. (2008). *Rcmdr: R Commander.* R package version 1.3–15.

Fox, J. & Anderson, R. (2005). Using the R statistical computing environment to teach social science statistics courses. Available: http://socserv.mcmaster.ca/jfox/Teaching-with-R.pdf and see also http://socserv.mcmaster.ca/jfox/Courses/R-course/index.html.

Galton, F. (1886). Regression towards mediocrity in hereditary stature. *Journal of the Anthropological Institute, 15*, 246–263.

Goldstein, H. (2003). *Multilevel statistical methods* (3rd ed.). London: Edward Arnold.

Groeneveld, R. A. & Meeden, G. (1984). Measuring skewness and kurtosis. *Statistician, 33*, 391–399.

Halvorsen, K. (2007). *ElemStatLearn: Data sets, functions and examples from the book: 'The Elements of Statistical Learning, Data Mining, Inference, and Prediction' by Trevor Hastie, Robert Tibshirani and Jerome Friedman.* R package version 0.1–4.

Hand, D. J. (1994). Deconstructing statistical questions. *Journal of the Royal Statistical Society: A, 157*, 317–356.

Harrell, F. E. Jr, & colleagues (2007). *Hmisc: Harrell miscellaneous*. R package version 3.4–3.

Hastie, T. (2008). *gam: Generalized additive models*. R package version 1.0.

Hastie, T. & Efron, B. (2007). *lars: Least angle regression, lasso and forward stagewise*. R package version 0.9–7.

Hastie, T. & Tibshirani, R. (1990). *Generalized additive models*. London: Chapman & Hall.

Hastie, T., Tibshirani, R. & Friedman, J. (2001). *The elements of statistical learning: Data mining, inference, and prediction*. New York: Springer.

Hill, C., Abraham, C. & Wright, D. B. (2007). Can theory-based messages in combination with cognitive prompts promote exercise in classroom settings? *Social Science & Medicine, 65*, 1049–1058.

Hinde, J. & Demétrio, C. G. B. (1998). Overdispersion: Models and estimation. *Computational Statistics & Data Analysis, 27*, 151–170.

Hoaglin, D. C., Mosteller, F. & Tukey, J. W. (eds) (2000). *Understanding robust and exploratory data analysis*. New York: John Wiley & Sons.

Hoffmann, J. P. (2004). *Generalized linear models: An applied approach*. Boston, MA: Pearson Education Inc.

Hoffman, L. & Rovine, M. J. (2007). Multilevel models for the experimental psychologist: Foundations and illustrative examples. *Behavior Research Methods, 39*, 101–117.

Holland, P. W. & Rubin, D. B. (1983). 'On Lord's paradox'. In Wainer, H. and Messick, S. (Eds), *Principals of modern psychological measurement*. Hillsdale, NJ: Lawrence Erlbaum Associates, pp. 3–35.

Hox, J. (2002). *Multilevel analysis: Techniques and applications*. London: Lawrence Erlbaum Associates.

Kreft, I. I. & de Leeuw, J. (1998). *Introducing multilevel modeling*. London: Sage Publications.

London, K., Bruck, M. & Melnyk, L. (in press). Persistence of facilitation and misinformation effects in event memory following a 10 month delay. *Law and Human Behavior*.

Lord, F. M. (1953). On the statistical treatment of football numbers. *American Psychologist, 8*, 750–751.

Lord, F. M. (1967). A paradox in the interpretation of group comparisons. *Psychological Bulletin, 72*, 304–305.

Lumley, T. (2006). *leaps: regression subset selection*. R package version 2.7. using Fortran code by Alan Miller.

MacCallum, R. C., Zhang, S., Preacher, K. J. & Rucker, D. D. (2002). On the practice of dichotomization of quantitative variables. *Psychological Methods, 7*, 19–40.

MacKinnon, D. P., Fairchild, A. J. & Fritz, M. S. (2007). Mediation analysis. *Annual Review of Psychology, 58*, 593–614.

Maronna, R. A., Martin, R. D. & Yohai, V. J. (2006). *Robust statistics: Theory and methods*. Chichester, UK: Wiley.

McCullagh, P. & Nelder, J. A. (1989). *Generalized linear models* (2nd ed.). London: Chapman and Hall.

Meehl, P. E. (1978). Theoretical risks and tabular asterisks: Karl, Ronald, and slow progress of soft psychology. *Journal of Consulting and Clinical Psychology, 46*, 806–834.

Micceri, T. (1989). The unicorn, the normal curve, and other improbable creatures. *Psychological Bulletin, 105*, 156–166.

Mosteller, F. & Tukey, J. W. (1977). *Data analysis and regression: A second course in statistics*. Reading, MA: Addison-Wesley Publishing Company.

Murrell, P. (2006). *R graphics*. Boca Raton, FL: Chapman & Hall/CRC.

Nelder, J. A. & Wedderburn, R. W. M. (1972). Generalized linear models. *Journal of the Royal Statistical Society: Series A, 135,* 370–384.

Park, M. Y. & Hastie, T. (2007a). An L1 regularization-path algorithm for generalized linear models. *Journal of the Royal Statistical Society: Series B, 69,* 659–677.

Park, M. Y. & Hastie, T. (2007b). *glmpath: L1 regularization path for generalized linear models and Cox proportional hazards model*. R package version 0.94.

Pineheiro, J. C. & Bates, D. M. (2000). *Mixed-effects models in S and S-Plus*. New York: Springer.

Pinheiro, J. C., Bates, D. M., DebRoy, S., Sarkar, D. & R Core team (2008). *nlme: Linear and nonlinear mixed effects models*. R package version 3.1–89.

Preacher, K. J. & Hayes, A. F. (2004). SPSS and SAS procedures for estimating indirect effects in simple mediation models. *Behavior Research Methods, Instruments, & Computers, 36,* 717–731.

R Development Core Team (2008). *R: A language and environment for statistical computing*. R Foundation for Statistical Computing, Vienna, Austria.

R core members, DebRoy, S., Bivand, R. & others (2008). *foreign: Read data stored by Minitab, S, SAS, SPSS, Stata, Systat, dBase,* R package version 0.8–26.

Roberts, L., Lafta, R., Garfield, R., Khudhairi, J. & Burnham, G. (2004). Mortality before and after the 2003 invasion of Iraq: cluster sample survey. *The Lancet, 364,* 1857–1864.

Robinson, W. S. (1950). Ecological correlations and the behavior of individuals. *American Sociological Review, 15,* 351–357.

Rubin, D. B. (1974). Estimating causal effects of treatments in randomized and nonrandomized studies. *Journal of Educational Psychology, 66,* 688–70.

Sarkar, D. (2008). *lattice: Lattice Graphics*. R package version 0.17–8.

Shumway, R. H. & Stoffer, D. S. (2006). *Time series analysis and its applications: With R examples* (2nd ed.). New York: Springer.

Siegel, S. (1956). *Nonparametric statistics for the behavioral sciences*. New York: McGraw-Hill Book Company.

Siegel, S. & Castellan, N. J. (1988). *Nonparametric statistics for the behavioral sciences* (2nd ed.). New York: McGraw-Hill Education.

Singer, J. D. & Willett, J. B. (2003). *Applied longitudinal data analysis: Modeling change and event occurrence*. New York: Oxford University Press.

Stigler, S. M. (1986). *The history of statistics: The measurement of uncertainty before 1900*. Cambridge, MA: Harvard University Press.

Stigler, S. M. (1999). 'Gauss and the invention of least squares'. In S. M. Stigler (Ed.) *Statistics on the table: The history of statistical concepts and methods*. Cambridge, MA: Harvard University Press, pp. 320–331. Originally published in *Annals of Statistics, 9,* 465–474.

Sullivan, L. M., Dukes, K. A. & Losina, E. (1999). An introduction to hierarchical linear modelling. *Statistics in Medicine, 18,* 855–888.

Thornton, S. (2007). *Financial inequality in higher education: The annual report on the economic status of the profession* 2006–2007. American Association of University Professors (AAUP). Available: http://www.aaup.org/NR/rdonlyres/B25BFE69-BCE7-4AC9-A644-7E84FF14B883/0/zreport.pdf.

Tibshirani, R. (1996). Regression shrinkage and selection via the Lasso. *Journal of the Royal Statistical Society Series: B, 58,* 267–288.

Tibshirani, R., Saunders, M., Rosset, S., Zhu, J. & Knight, K. (2005). Sparsity and smoothness via the fused lasso. *Journal of the Royal Statistical Society Series: B, 67,* 91–108.

Tufte, E. R. (2001). *The visual display of quantitative information* (2nd ed.). Cheshire, Conn.: Graphics Press.

Tukey, J. W. (1960). A survey of sampling from contaminated normal distributions. In Olkin, I., Ghurye, S., Hoeffding, W., Madow, W. and Mann, H. (Eds), *Contributions to probability and statistics*. Stanford, CA: Stanford University Press, pp. 448–485.

Tukey, J. W. (1977). *Exploratory data analysis*. Reading, Mass.: Addison-Wesley.

UNICEF (2007). Report card on child well-being in rich countries. Innocenti Report Card 7. Available: http://www.unicef-icdc.org/.

Velleman, P. & Wilkinson, L. (1993). Nominal, ordinal, interval, and ratio typologies are misleading for classifying statistical methodology. *American Statistician, 47*, 65–72.

Venables, W. N. & Ripley, B. D. (2002). *Modern applied statistics with S/S-Plus* (4th ed.). New York: Springer.

Vrij, A. (2005). Criteria-based content analysis: A qualitative review of the first 37 studies. *Psychology, Public Policy, & Law, 11*, 3–41.

Wachsmuth, A., Wilkinson, L. & Dallal, G. E. (2003). Galton's bend: An undiscovered nonlinearity in Galton's family stature regression data and a likely explanation based on Pearson and Lee's stature data. *American Statistician, 57*, 190–192.

Wainer, H. (1984). How to display data badly. *American Statistician, 38*, 137–147.

Wainer, H. (1991). Adjusting for differential base rates: Lord's paradox again. *Psychological Bulletin, 109*, 147–151.

Wainer, H. & Brown, L. M. (2004). Two statistical paradoxes in the interpretation of group differences: Illustrated with Medical School Admission and Licensing Data. *American Statistician, 58*, 117–123.

Wand, M. P. (1997). Data-based choice of histogram bin width. *American Statistician, 51*, 59–64.

Wehrens, R. & Mevik, B.-H. (2007). *pls: Partial Least Squares Regression (PLSR) and Principal Component Regression (PCR)*. R package version 2.1–0.

Wilcox, R. R. (1995). ANOVA: A paradigm for low power and misleading measures of effect size. *Review of Educational Research, 65*, 51–77.

Wilcox, R. R. (1998). How many discoveries have been lost by ignoring modern statistical methods? *American Psychologist, 53*, 300–314.

Wilcox, R. R. (2003a). *Applying contemporary statistical techniques*. Orlando, FL: Academic Press.

Wilcox, R. R. (2003b). Inferences based on multiple skipped correlations. *Computational Statistics & Data Analysis, 44*, 223–236.

Wilkinson, G. & Rogers, C. (1973). Symbolic description of factorial models for the analysis of variance. *Applied Statistics, 22*, 392–399.

Wilkinson, L. & the Task Force on Statistical Inference, APA Board of Scientific Affairs (1999). Statistical methods in psychology journals: Guidelines and explanations. *American Psychologist, 54*, 594–604.

Wood, S. N. (2006). *Generalized additive models: An introduction with R*. Boca Raton, FL: Chapman & Hall/CRC.

Wright, D. B. (1997). Extra-binomial variation in multilevel logistic models with sparse structures. *British Journal of Mathematical and Statistical Psychology, 50*, 21–29.

Wright, D. B. (1998). Modelling clustered data in autobiographical memory research: The multilevel approach. *Applied Cognitive Psychology, 12*, 339–357.

Wright, D. B. (2003). Making friends with your data: Improving how statistics are conducted and reported. *British Journal of Educational Psychology, 73*, 123–136.

Wright, D. B. (2006a). 'The art of statistics: A survey of modern statistics'. In Alexander, P. A. & Winne, P. H. (Eds), *Handbook of educational psychology* (2nd ed.). Mahwah, NJ: Erlbaum, pp. 879–901.

Wright, D. B. (2006b). Comparing groups in a before-after design: When *t*-test and ANCOVA produce different results. *British Journal of Educational Psychology, 76*, 663–675.

Wright, D. B. (2008). A new improved ANCOVA. *Psychologist, 21*, 225–226.

Wright, D. B., Boyd, C. E. & Tredoux, C. G. (2003). Inter-racial contact and the own race bias for face recognition in South Africa and England. *Applied Cognitive Psychology, 17*, 365–373.

Wright, D. B. & Hall, M. (2007). How a 'Reasonable Doubt' instruction affects decisions of guilt. *Basic and Applied Social Psychology, 29*, 85–92.

Wright, D. B., Horry, R. & Skagerberg, E. M. (in press). Functions for traditional and multilevel approaches to signal detection theory. *Behavior Research Methods*.

Wright, D. B. & Livingston-Raper, D. (2001). Memory distortion and dissociation: Exploring the relationship in a non-clinical sample. *Journal of Trauma and Dissociation, 3*, 97–109.

Wright, D. B. & London, K. (2009). *First (and second) steps in statistics* (2nd ed.). London: Sage.

Wright, D. B. & London, K. (in press). Multilevel modelling: Beyond the basic applications. *British Journal of Mathematical and Statistical Psychology*.

Wright, D. B. & Williams, S. (2003). Producing bad results sections. *Psychologist, 16*, 644–648.

Zhao, P. & Yu, B. (2006). On model selection consistency of Lasso. *Journal of Machine Learning Research, 7*, 2541–2563.

Index

Note: Entries in bold are R functions. The suffix 'n' following locators refers to footnotes and 'f' refers to figures.

Supporting researchers for more than forty years

Research methods have always been at the core of SAGE's publishing. Sara Miller McCune founded SAGE in 1965 and soon after, she published SAGE's first methods book, Public Policy Evaluation. A few years later, she launched the Quantitative Applications in the Social Sciences series – affectionately known as the "little green books".

Always at the forefront of developing and supporting new approaches in methods, SAGE published early groundbreaking texts and journals in the fields of qualitative methods and evaluation.

Today, more than forty years and two million little green books later, SAGE continues to push the boundaries with a growing list of more than 1,200 research methods books, journals, and reference works across the social, behavioral, and health sciences.

From qualitative, quantitative, mixed methods to evaluation, SAGE is the essential resource for academics and practitioners looking for the latest methods by leading scholars.

www.sagepublications.com

The Qualitative Research Kit

Edited by Uwe Flick

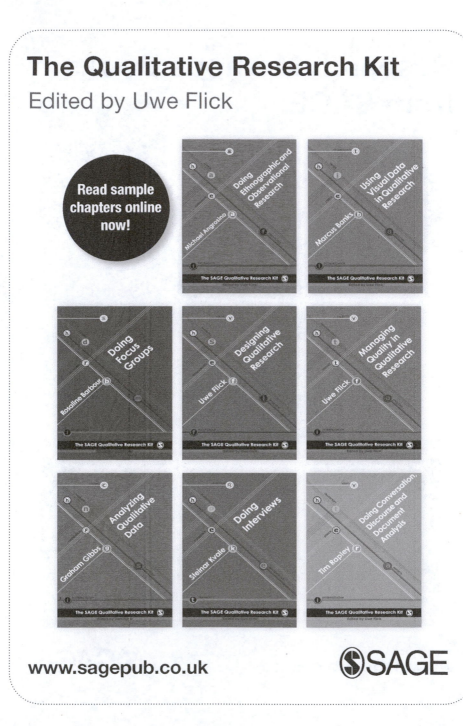

Read sample chapters online now!

Doing Ethnographic and Observational Research — Michael Angrosino

Using Visual Data in Qualitative Research — Marcus Banks

Doing Focus Groups — Rosaline Barbour

Designing Qualitative Research — Uwe Flick

Managing Quality in Qualitative Research — Uwe Flick

Analyzing Qualitative Data — Graham Gibbs

Doing Interviews — Steinar Kvale

Doing Conversation, Discourse and Document Analysis — Tim Rapley

The SAGE Qualitative Research Kit

www.sagepub.co.uk

SAGE